Capillary wetting of heterogeneous powders

Capillary wetting of heterogeneous powders

Vom Promotionsausschuss der
Technischen Universität Hamburg

zur Erlangung des akademischen Grades

Doktor-Ingenieur (Dr.-Ing.)

genehmigte Dissertation

von

Jana Christina Kammerhofer

aus

Burghausen

2019

Gutachter:

1. Prof. Dr.-Ing. habil. Dr. h.c. Stefan Heinrich
2. Prof. Dr.-Ing. habil. Irina Smirnova
3. Prof. Dr.-Ing. habil. Stefan Palzer

Tag der mündlichen Prüfung:

13.06.2019

Bibliografische Information der Deutschen Nationalbibliothek

Die Deutsche Nationalbibliothek verzeichnet diese Publikation in der Deutschen Nationalbibliographie; detaillierte bibliographische Daten sind im Internet über http://dnb.d-nb.de abrufbar.

1. Aufl. - Göttingen: Cuvillier, 2019

   Zugl.: (TU) Hamburg, Univ., Diss., 2019

© CUVILLIER VERLAG, Göttingen 2019
   Nonnenstieg 8, 37075 Göttingen
   Telefon: 0551-54724-0
   Telefax: 0551-54724-21
   www.cuvillier.de

  ISBN 978-3-7369-7082-3
  eISBN 978-3-7369-6082-4

Preface

An erster Stelle möchte ich mich bei meinem Doktorvater, Herrn Prof. Dr.-Ing. habil. Dr. h.c. Stefan Heinrich, Leiter des Instituts für Feststoffverfahrenstechnik und Partikeltechnologie, für seine Unterstützung und Förderung bedanken. Zu jeder Zeit hat mir Herr Heinrich vollstes Vertrauen entgegengebracht, was ich als besondere Motivation empfunden habe. Darüber hinaus möchte ich mich dafür bedanken, dass er mir die Teilnahme an nationalen und internationalen Konferenzen ermöglicht hat. Das alles hat zum Gelingen dieser Arbeit beigetragen. Auch der Zweitprüfering Frau Prof. Dr.-Ing. habil. Irina Smirnova, Leiterin des Instituts für Thermische Verfahrenstechnik, und dem Prüfungsvorsitzenden Prof. Dr.-Ing. habil. Raimund Horn, Leiter des Instituts für Chemische Reaktionstechnik, gilt mein besonderer Dank für die Übernahme des jeweiligen Amtes.

Ebenso danke ich Herrn Prof. Dr.-Ing. habil. Stefan Palzer, Executive Vice President und Chief Technology Officer von Nestlé, für sein großes Interesse an meiner Forschung, das entgegengebrachte Vertrauen, den fachlichen Rat und die Übernahme des externen Gutachtens.

Weiterhin möchte ich mich ganz besonders bei Dr.-Ing. Lennart Fries, Dr.-Ing. Laurent Forny und Dr.-Ing. Julien Dupas bedanken, die als externe Betreuer vom Nestlé Research Center in Lausanne und vom Nestlé Product Technology Center in Orbe meine Doktorarbeit begleitet und unterstützt haben. Die zahlreichen fachlichen Diskussionen während unserer monatlichen Telefonate haben mir stets wertvolle Anregungen für meine Forschung gegeben, was maßgeblich zum erfolgreichen Abschluss dieser Doktorarbeit beigetragen hat.

Mein Dank gilt auch allen meinen Kolleginnen und Kollegen am Institut für Feststoffverfahrenstechnik und Partikeltechnologie für ihre Hilfsbereitschaft und das tolle Arbeitsklima. Eine große Hilfe waren alle Studentinnen und Studenten, die mit der Anfertigung ihrer Bachelor-, Master- und Projektarbeiten oder als Hiwi zum Gelingen meiner Doktorarbeit beigetragen haben. Ganz herzlich bedanke ich mich auch bei Dr.-Ing. Swantje Pietsch, mit der ich mir vier Jahre das Büro teilen durfte und die meine Zeit am SPE ganz tief geprägt hat. Danke für den heiteren Büroalltag, die unvergesslichen Dienstreisen und die Antworten auf alle meine Fragen.

Darüber hinaus möchte ich noch einen ganz besonderen Dank an meine Eltern Margret und Peter Kammerhofer aussprechen, die mich während meiner gesamten Ausbildungszeit zu 100% unterstützt haben und mir stets uneingeschränktes Vertrauen entgegengebracht haben. Last but not least bedanke ich mich bei meinem Freund Cornelius Zaubitzer, der als hervorragende Quelle für Motivation und Energie während der Anfertigung meiner Doktorarbeit gedient hat und immer an mich glaubt.

Lausanne, 27.08.2019
Jana Kammerhofer

Abstract

Capillary wetting of heterogeneous powders

by Jana Christina KAMMERHOFER

HAMBURG UNIVERSITY OF TECHNOLOGY

Institute of Solids Process Engineering and Particle Technology

The wetting kinetics of liquid-powder systems is an important factor with regard to many practical applications, such as coating, granulation, agglomeration or powder reconstitution. Wetting as the first step during food powder reconstitution strongly depends on the interactions of the wetting liquid and the particle surface, expressed by the contact angle. Due to the heterogeneous composition of food materials including hydrophilic and hydrophobic surfaces, also the wetting process is of heterogeneous nature. Furthermore, the solubility of food ingredients, such as sugars, increases the complexity of understanding and describing the wettability.

The objective of this thesis is the enhancement of understanding and prediction of capillary wetting of heterogeneous, soluble food powders. Therefore, the capillary penetration was studied experimentally by increasing stepwise the complexity. Firstly, the different influencing factors on capillary wetting were studied separately. The heterogeneity in terms of contact angle and the effect of solubility during capillary wetting as the two main factors were investigated and combined afterwards. Furthermore, a new model based on physical equations was developed for capillary penetration into heterogeneous and soluble food systems. It was found that the viscosity increase is dominant during liquid penetration for the powder systems including the most soluble component (sucrose) at a high concentration. For powder mixtures containing a less soluble component with less viscosity development (sodium chloride), the hydrophobic contact angle drives the liquid penetration. The model is suitable for the prediction of capillary penetration into heterogeneous systems as long as the viscosity increase is the dominant effect. A further improvement of the model is required for enhancing the predition of systems where the change of pore network properties is the driving factor. As last step, the model was tested for a real food powder exhibiting heterogeneity in terms of composition and solubility. Even if the modelled results are already in a fair agreement with the experiments, the slight overestimation of the model could be still improved by integration of changing pore network properties in the model.

Contents

List of Figures

List of Tables

Symbols

Latin symbols

a	mass tranfer area	m^2
A	cross sectional area	m^2
Ar	Archimedes number	-
B	permeability	m^2
C	penetration rate	$\mathrm{kg/m}^2/\sqrt{\mathrm{s}}$
Ca	capillary number	-
d	diameter of particle	m
D	diffusion coefficient	$\mathrm{m}^2\,\mathrm{s}^{-1}$
D_{cap}	characteristic diameter for capillary pressure	-
D_{vis}	characteristic diameter for viscous drag	-
e	defect ratio	-
E	surface energy	$\mathrm{N\,m}^{-2}$
f	area surface fraction	-
F	force	N
f_a	form factor	-
g	gravitational acceleration	$\mathrm{m\,s}^{-2}$
G	experimental constant	-
h	height	m
H	height of powder bed	m
j	factor depending on particle size, shape, porosity	-
k	constant or rate	-
K	capillary constant	mm^5
k_B	Boltzmann constant	$\mathrm{J\,K}^{-1}$
l	length or distance	m
L	lengthscale	-
M	mass	kg
P	pressure	$\mathrm{N\,m}^{-2}$

r	radius	m
R	roughness parameter	-
R^2	coefficient of determination	-
R_0	hydrodynamic radius	m
R_a	specific roughness parameter	m
Re	Reynolds number	-
Q	flow through capillary	$\mathrm{m^3\,s^{-1}}$
Q_3	volume related cumulative distribution	%
S_v	volume specific surface area	$\mathrm{m^2\,m^{-3}}$
Sc	Schmidt number	-
Sh	Sherwood number	-
Sp	spreading parameter	-
t	time	s
u	velocity	$\mathrm{m\,s^{-1}}$
V	volume	$\mathrm{m^3}$
w	width	m
W	Lambert function	-
x	position in x-axis	m
X	load of solution	$\mathrm{kg_s/kg_l}$
z	position in z-axis	m
z_c	distance betw. initial and depressed liquid surface	m

Greek symbols

β	mass transfer coefficient	$\mathrm{m\,s^{-1}}$
δ	roundness	-
ε	porosity	-
ζ	proportion of hydrophobic surface	-
η	dynamic viscosity	$\mathrm{Pa\,s}$
θ	contact angle of liquid on solid	°
ρ	density	$\mathrm{kg\,m^{-3}}$
σ	surface tension	$\mathrm{N\,m^{-1}}$
τ	tortuosity	-
ϕ	angle between horizontal and s-l-g contact locus	°
χ	angle of inclined capillary	°
ψ	sphericity	-

Indices

$*$	critical or modified
0	initial
1	component 1
$10,3$	10% of the total volume
2	component 2
32	Sauter
50	mean
$50,3$	50% of the total volume
$90,3$	90% of the total volume
a	apparent
ad	advancing
cap	capillary
CB	Cassie-Baxter
$crit$	critical
d	dynamic
$diss$	dissolution
dr	drop
eff	effective
eq	equilibrium
exp	experimental
g	gas
gr	gravity
h	height depending
hex	hexane
i	intrinsic
in	inertial
l	liquid
lam	laminar
m	mass related
p	pore
rec	receding
s	solid
sat	saturation
tap	tapped
$theo$	theoretical
$turb$	turbulent
vis	viscous

v	volume related
wet	wetting
w	water

Abbreviations

exp	experiment
ff	free fat
SEM	scanning electron microscope
WMP	whole milk powder

1

Introduction

1.1 Wetting fundamentals

The wetting of solid surfaces by a liquid plays a considerable role in biology, daily life and industry (Marmur, 1992b). In many processes and applications, which are applied in chemical, metallurgy, ceramic, petrochemical, pharmaceutical and food industry, the wettability decides about the quality or the success of the outcome. For instance, in the field of solids process engineering, the wettability of a particulate system is an important step during agglomeration, granulation or coating processes which can decide about the product quality (Charles-Williams et al., 2013, Hapgood et al., 2003, Iveson et al., 2001). Hence, an overview about fundamentals of wetting of a flat surface, a single capillary and a porous system is provided in this chapter.

1.1.1 Droplet spreading

When a liquid droplet gets in contact with a flat solid surface, two cases can be distinguished: total wetting and partial wetting. Sometimes even a third case is described, the partial nonwetting (Masoodi and Pillai, 2012). The three different wetting cases are presented in Figure 1.1.

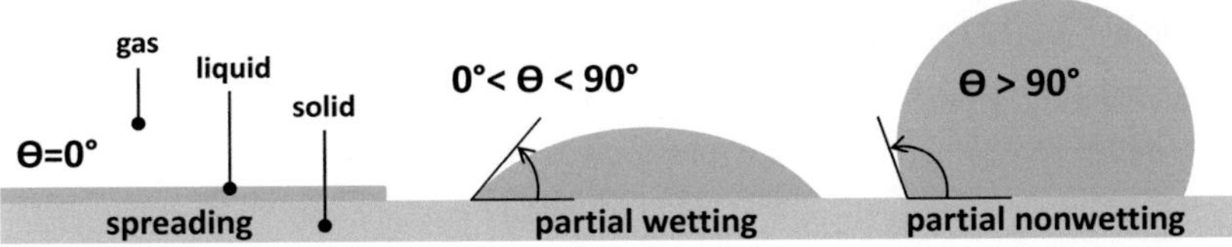

FIGURE 1.1: Three wetting cases: spreading, partial wetting and partial nonwetting.

In order to characterize the wettability, the spreading parameter Sp, which measures the difference between the surface energies of the substrate in a dry E_{dry} and in a wetted state E_{wet} (Eq. 1.1), is consulted. It can also be expressed by the surface tensions at the solid/air $\sigma_{s,g}$, solid/liquid $\sigma_{s,l}$ and liquid/air interfaces σ (Eq. 1.2) (de Gennes et al., 2004):

$$Sp = E_{dry} - E_{wet}, \tag{1.1}$$

$$Sp = \sigma_{s,g} - (\sigma_{s,l} + \sigma). \tag{1.2}$$

$Sp > 0$ means that the liquid spreads over the solid surface due to a strong affinity and to lower its surface energy. The resulting contact angle θ which forms at the three-phase boundary point is zero. The partial wetting is expressed by $Sp < 0$ and $\theta < 90°$ whereas the partial nonwetting is defined as $Sp < 0$ and $\theta > 90°$. Thus, besides the spreading parameter, the contact angle provides an opportunity to express the interactions between a solid and a liquid. Young (1805) defined the intrinsic contact angle on a molecular level which is based on a force balance (Eq. 1.3):

$$\sigma \cdot \cos\theta = \sigma_{s,g} - \sigma_{s,l}. \tag{1.3}$$

Since this contact angle can be only measured on ideal, flat, smooth and homogeneous surfaces, the apparent contact angle is observable on a microscopic level due to surface imperfections, reactions between fluid and solid material, roughness or chemical heterogeneity (Marmur, 1992a, Palzer et al., 2001, Reinke et al., 2015). These chemical and physical irregularities form non-ideal surfaces resulting in the fact that the static contact angle is not longer unique for the dynamic case (de Gennes et al., 2004, Li and Neumann, 1992). When a liquid droplet moves along a real solid surface, the contact angle varies between a maximum and a minimum value, which are termed as advancing θ_{ad} and receding contact angel θ_{rec}, respectively. The advancing contact angle is larger than the static value and occurs when a droplet is inflated and advances across a dry solid surface. On the contrary, when deflating a droplet the liquid recedes from the surface resulting in the smaller receding contact angle (de Gennes et al., 2004, Li and Neumann, 1992). The difference between the maximum and minimum contact angles is called the contact angle hysteresis. On clean and smooth surfaces, the hysteresis can be smaller than $5°$, while rough and dirty surfaces can extend the hysteresis to more than $50°$ (de Gennes et al., 2004). The dynamic contact angle depends on the velocity of the contact line, but is indenpendent of the fact if it is a two-dimensional meniscus or a axisymmetric droplet (Katoh et al., 2010, 2015). In literature, different correlations are available to describe the velocity dependence of the contact angle. Cox (1986) derived the following equation (Eq. 1.4), where θ_d is the dynamic contact angle, G is an experimental constant, Ca is the capillary number, L is the typical lengthscale of the system, L_{slip} is the slip length, η is the viscosity, u is the velocity and σ is the surface tension:

$$\theta_d = \left(\theta^3 + 9 \cdot G \cdot Ca\right)^{1/3}, \tag{1.4}$$

with

$$G = \ln\left(\frac{L}{L_{slip}}\right),$$
$$Ca = \frac{\eta \cdot u}{\sigma}.$$

Katoh et al. (2010) presented another approach (Eq.1.5) to express the dynamic contact angle, where e is the ratio occupied by defects:

$$|\cos\theta_d - \cos\theta| = \frac{3 \cdot (1-e)}{e \cdot \tan\theta_d} \cdot Ca. \tag{1.5}$$

As already mentioned above, surface heterogeneities are generally divided into two groups, physical and chemical heterogeneities. Contact angles on chemically heterogeneous, but smooth surfaces can be expressed by an approach presented by Cassie and Baxter (1944). Assuming the surface consists of two components, component 1 and 2, Figure 1.2 presents the composition of the Cassie-Baxter contact angle.

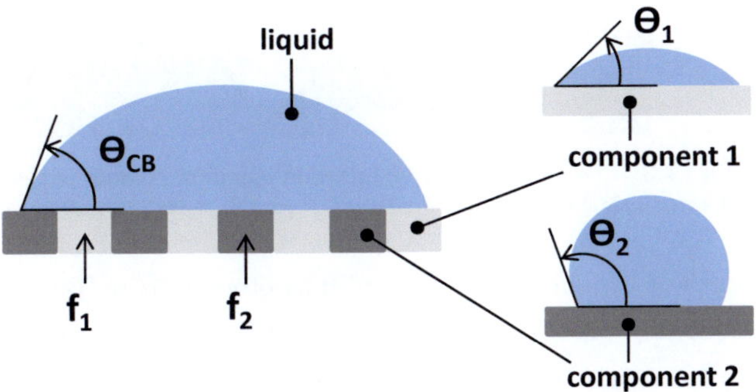

FIGURE 1.2: Cassie-Baxter contact angle θ_{CB} on chemical heterogeneous surface consisting of component 1 with θ_1 and component 2 with θ_2.

Eq.(1.6) can be applied to calculate the Cassie-Baxter contact angle θ_{CB} of two components. θ_1 and f_1 are the contact angle and the area surface fraction of component 1 and θ_2 and f_2 are the contact angle and the area surface fraction of component 2. The angle θ_{CB} appears as the apparent contact angle on the heterogeneous surface:

$$\cos\theta_{CB} = f_1 \cdot \cos\theta_1 + f_2 \cdot \cos\theta_2. \tag{1.6}$$

Physical heterogeneity on surfaces occurs due to geometric roughness which also influences the contact angle. This phenomenon of contact angles on rough, but chemical homogeneous surfaces was explained by Wenzel (1936, 1949). Eq. (1.7) presents the relation between the apparent contact angle, the roughness parameter R and the intrinsic contact angle:

$$\cos\theta_a = R \cdot \cos\theta_i, \tag{1.7}$$

with

$$R = \frac{\text{actual surface area}}{\text{geometric surface area}}.$$

Wenzel's relation describes the effect that roughness decreases the apparent contact angles if it is below 90° and increases the apparent contact angle if it is above 90° which can be seen in Figure 1.3.

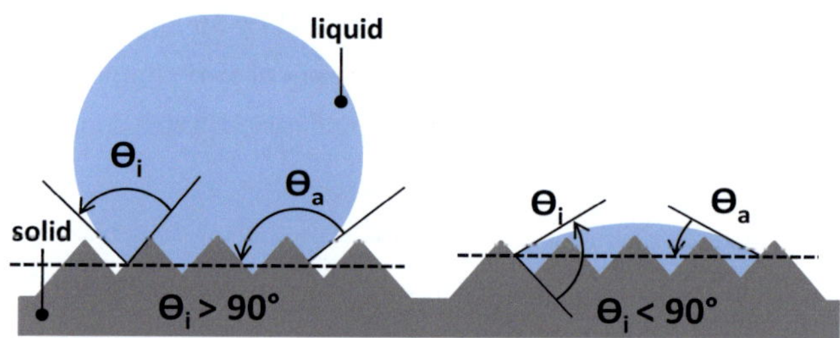

FIGURE 1.3: Effect of surface roughness on apparent contact angle θ_a for the two cases for the intrinsic contact angle θ_i: $\theta_i > 90°$ and $\theta_i > 90°$, modified from (Dullien, 1992).

This effect was experimentally confirmed by other authors (Busscher et al., 1984, Palzer et al., 2001) for the static case. Furthermore, Busscher et al. (1984) figured out that there is no influence of surface roughness on the contact angle as long as the roughness is below 0.1 µm. For the dynamic case, Palzer et al. (2001) generally observed higher contact angles with an increased surface roughness independent of the magnitude of the contact angle.

1.1.2   Liquid imbibition into single pore

Capillarity describes the behaviour of liquids in small tubes, called capillaries, due to interfacial forces. The capillary pressure P_{cap} is the driving force which pulls the liquid into the tubes (Masoodi and Pillai, 2012). r_p is the radius of the pore. Generally, it can be distinguished between two cases. If adhesion forces between liquid and capillary wall are larger than cohesion forces within the liquid, capillary rise can be observed. In the opposite case, capillary depression occurs (Dobrinski et al., 2010). A critical contact angle of 90° is the limiting case. Solid-liquid systems, which form contact angles below 90° result in capillary rise:

$$P_{cap} = \frac{2 \cdot \sigma \cos \theta}{r_p}. \tag{1.8}$$

During the penetration of liquid into a single capillary, different forces act in different directions and decide about the dynamics of the rise. The overall force balance is described by several authors in literature (Marmur, 1992a, Martic et al., 2002, Zhmud et al., 2000) and includes the capillary force F_{cap}, the viscous force F_{vis}, the gravity force F_{gr} and the inertial force F_{in} (Eq.(1.9):

$$F_{cap} = F_{vis} + F_{gr} + F_{in}. \tag{1.9}$$

Converting the force balance into a pressure balance by relating the forces to the cross-sectional area of the capillary or pore, Eq. (1.10) is obtained. The equation consists of four pressure terms, the capillary pressure, the hydrostatic pressure, the viscous pressure loss which is expressed in the Hagen-Poiseuille equation and the inertia term. r_p is the radius of the capillary or pore, ρ_l is the density of the liquid, g is the gravitational acceleration, h is the height and t is the time. In order to describe the rise into inclined capillaries as well, Fries and Dreyer (2008a) added the term $\sin \chi$ to the hydrostatic pressure where χ is the angle which forms between inclined capillary and free liquid surface:

$$\frac{2 \cdot \sigma \cdot \cos \theta}{r_p} = \frac{8 \cdot \eta \cdot h}{r_p^2} \cdot \frac{dh}{dt} + \rho_l \cdot g \cdot h \cdot \sin \chi + \rho_l \cdot \left[h \cdot \frac{d^2 h}{dt^2} + \left(\frac{dh}{dt} \right)^2 \right]. \tag{1.10}$$

During the penetration of a liquid into a capillary, different effects are dominant depending on the time stage. Four cases can be distinguished (Fries and Dreyer, 2008b): the purely inertial time stage, the visco-inertial time stage, the purely viscous time stage and the viscous and gravitational time stage. A final stage is added, which describes the equilibrium state of liquid in a capillary.

Inertial time stage

Quéré (1997) presented the following approach for the very first moments when the liquid gets in contact with the capillary. During this time stage, the viscous and the gravity term can be neglected. Eq. (1.11) is the simplified version of Eq. (1.10):

$$\frac{2 \cdot \sigma \cdot \cos\theta}{r_p \cdot \rho_l} = h \cdot \frac{d^2 h}{dt^2} + \left(\frac{dh}{dt}\right)^2 .$$ (1.11)

The differential equation was solved by Quéré (1997) leading to a linear law for the meniscus height in a capillary versus time (Eq. 1.12):

$$h = t \cdot \sqrt{\frac{2 \cdot \sigma \cdot \cos\theta}{\rho_l \cdot r_p}} .$$ (1.12)

Visco-inertial time stage

Is the penetration dominated by viscous and inertial forces, Bosanquet (1923) derived the following simplification to express the behaviour of the flow (Eq.1.13):

$$h^2 = \frac{2b}{a} \cdot \left[t - \frac{1}{a} \cdot \left(1 - e^{-at}\right) \right],$$ (1.13)

with

$$a = \frac{8 \cdot \eta}{r_p^2 \cdot \rho_l},$$
$$b = \frac{2 \cdot \sigma \cdot \cos\theta}{r_p \cdot \rho_l} .$$

Eq. (1.13) transforms into the Washburn equation for $t \to \infty$. Ichikawa and Satoda (1994) studied the same case, but expressed the equation in a dimensionless form.

Viscous time stage

The viscous time stage during capillary rise is the most known and discussed case in literature. Neglecting the inertial and gravity forces Eq. (1.10) can be simplified and written as:

$$\frac{2 \cdot \sigma \cdot \cos \theta}{r_p} = \frac{8 \cdot \eta \cdot h}{r_p^2} \cdot \frac{dh}{dt}.$$ (1.14)

Solving Eq. (1.14) by inserting the initial condition $h(t = 0) = 0$ results in Eq. (1.15). This equation was derived by Washburn (1921) and Lucas (1918) independently from each other. But in the following the equation is titled as the Washburn equation:

$$h(t) = \sqrt{\frac{r_p \cdot \sigma \cdot \cos \theta}{2 \cdot \eta} \cdot t}.$$ (1.15)

The Washburn equation provides a basis for many modelling approaches of capillary rise phenomena in literature. This aspect is discussed more detailed in subsection 1.3.1.

Viscous and gravitational time stage

In the fourth stage, the penetration flow starts to be influenced by the gravity. Fries and Dreyer (2008a) figured out that the critical height when gravity has to be considered is at $h > 0.1 \cdot h_{eq}$, where h_{eq} is the equilibrium height in a capillary. Only the inertia effects can be neglected during this stage. An analytical solution is provided by using the Lambert function W for mathematical rearrangement (Eq. 1.16). a and b are constant factors including all solid and liquid properties and the gravity:

$$h(t) = \frac{a}{b} \left[1 + W \left(-e^{-1 - \frac{b^2 \cdot t}{a}} \right) \right].$$ (1.16)

Equilibrium stage

The equilibrium state is reached when the capillary pressure is balanced by the hydrostatic pressure (Eq. 1.18):

$$\frac{2 \cdot \sigma \cdot \cos \theta}{r_p} = \rho_l \cdot g \cdot h \cdot \sin \psi.$$ (1.17)

Assuming a vertical capillary ($\psi = 0$), the equilibrium height can be calculated with the following equation:

$$h_{eq} = \frac{2 \cdot \sigma \cdot \cos \theta}{\rho_l \cdot g \cdot r_p}.$$ (1.18)

Several other approaches are available in literature dealing for instance with short and long term solutions for the prediction of capillary rise (Chebbi, 2007, Zhmud et al., 2000) or with the influence of dynamic contact angles on the penetration behaviour (Chebbi, 2007, Hamraoui et al., 2000, Siebold et al., 2000).

Another influencing factor is the shape of the pore. Most studies assume cylindrical pores in which the liquid is imbibed. Birdi et al. (1988) and Wu et al. (2016) derived empirical equations for calculating the equilibrium height in rectangular capillaries. While the equation of Birdi et al. (1988) can be only applied for solid-liquid systems forming a contact angle of 0, Wu et al. (2016) included a term for the contact angle. Liquid penetration in angular gaps was studied by Bico and Quere (2002) for homogeneous square tubes and by O'Brien et al. (1968) for dissimilar walls with an angle.

O'Brien et al. (1968) developed a mathematical model to predict the capillary penetration between dissimilar plates (Eq. 1.19). The equation was evaluated with experiments which were conducted for different liquids between heterogeneous systems, such as glass-Teflon and glass-acrylic resin. The advancing contact angles of the materials were measured to be 14°, 74° and 110° for glass, acrylic and Teflon, respectively:

$$h_{eq} = \frac{\sigma \cdot (\cos \theta_1 + \cos \theta_2)}{\rho_l \cdot g \cdot w}. \tag{1.19}$$

In order to compare their measured data with the modelled rise, they used the factor $w \cdot h_{eq}$, which is the product of the gap width w multiplied by the equilibrium height. The deviation of the mean observed values from the predicted values is 2 % for the glass-acrylic system and 21 % for the glass-Teflon system.

1.1.3 Penetration into pore network

Capillary penetration into a pore network within a powder bed is based on the same physical fundamentals as the wicking into a single pore, however, the shape and the orientation of pores changes dramatically in these pore networks. Thus, significant differences in the behaviour are the consequences. Raux et al. (2013) compared the wicking of water into a capillary glass tube and into a porous media made of glass beads. For three different wettabilities, the glass tubes and beads were treated equally to modify the surface resulting in contact angles of 35°, 70° and 105°. For the contact angle of 35°, a wicking could be observed into both systems and for the angle of 105°, no wicking occurred. While the water penetrated into the 70°-glass tube, no penetration happened in the bed of glass beads, even if the contact angle was below 90°. This phenomenon was explained by geometrical reasons, since the pores within a powder bed are not cylindrical.

Nevertheless, a pore network within a powder bed is often described as a bundle of parallel cylinders with identical pore radii in literature which is also attributable to Washburn (1921). Therefore, in order to apply the Washburn equation for porous systems the pore radius r_p is replaced by an effective radius $r_{p,eff}$ of the parallel, cylindrical pores in Eq.(1.15). The following assumptions have to be made: 1. stationary, laminar flow, 2. no external pressure, 3. neglection of gravity force, 4. neglection of inertial force, 5. no-slip condition at the wall, 6. constant capillary pressure. Figure 1.4 presents a real powder bed and its model system.

The pore network can also be expressed as the product $k \cdot r_p$, where k is a constant to describe the randomly oriented capillaries (Bruil and van Aartsen, 1974). This product is determined by passing an ideal spreading liquid ($\theta = 0$) with known liquid parameters σ and η through the powder. Assuming a constant bed independent of the passing liquid this method is applicable (Rosen, 1978).

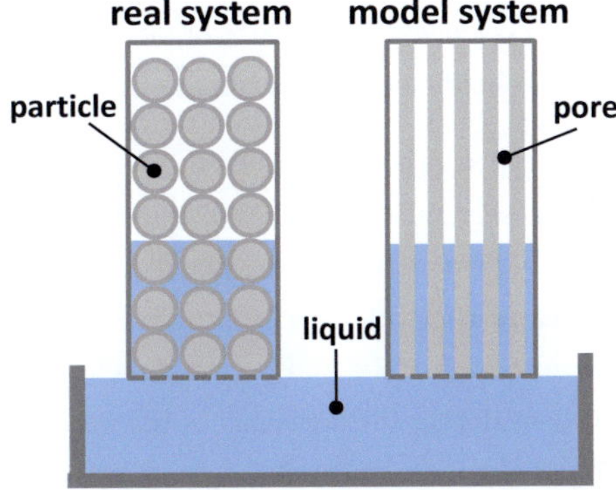

FIGURE 1.4: Real pore network (left) and model system (right) assuming a bundel of parallel, cylindrical pores (modified from Palzer (2000)).

Since the measurement of the liquid height for the Washburn approach is challenging, Murata and Naka (1983) and Kilau and Pahlman (1987) replaced the height by the penetration weight M_l of the liquid and used a modified equation (Eq. 1.20). An advantage of this modification is that the weight of the liquid uptake within the tube can be measured precisely, while the height of the liquid front is only visible at the glass wall of the tube but not in the interior. Therefore, the weight of the liquid within the tube is expressed by the height, the porosity ε of the powder bed, the cross sectional area of the tube A and the liquid density:

$$M(t) = \varepsilon \cdot A \cdot \rho_l \cdot \sqrt{\frac{k \cdot r_p \cdot \sigma \cos\theta}{2 \cdot \eta}} \cdot t. \tag{1.20}$$

The mass related Washburn equation can be also formulated differently (Eq. 1.21). Siebold et al. (1997) summarized the pore network parameters r_p, ε, A in the geometric factor K which is also obtained by using an ideal spreading liquid forming a contact angle of 0 on the solid. K is also called the capillary constant:

$$M(t) = \sqrt{\frac{K \cdot \rho_l^2 \cdot \sigma \cos\theta}{\eta}} \cdot t, \tag{1.21}$$

with

$$K = \frac{r_{p,eff} \cdot A^2 \cdot \varepsilon^2}{2}. \tag{1.22}$$

A critical step for this method is the creation of reproducible powder packings for the determination of K and the actual penetration experiment using the desired wetting liquid (Galet et al., 2010). Secondly, the correct choice of an ideal wetting liquid is decisive for this method (Prestidge and Ralston, 1995). Commonly used ideal wetting liquids are hexane, cyclohexane, heptane or octane (Chau, 2009, Galet et al., 2010, Iveson et al., 2000, Siebold et al., 1997, Susana et al., 2012).

Palzer (2000) presented another modified Washburn approach to describe the capillary wetting into a porous system. He suggested to use mercury porosimetry to get information about the real pore structure in the powder. Eq. (1.23) can be derived using the mean pore radius $r_{p,50}$ from the pore size distribution and introducing a shape factor ψ. This shape factor converts the real pore radius into an equivalent pore radius of the porous system by considering the deviation of the pore shape from the cylindrical shape and the alteration of the pore length due to curvature:

$$h(t) = \sqrt{\frac{r_{p,50} \cdot \psi \cdot \sigma \cdot \cos\theta}{2 \cdot \eta}} \cdot t. \tag{1.23}$$

If there is no possibility to measure the pore size distribution of a powder system, two further alternatives to determine pore radii in powder beds are proposed by Hapgood et al. (2002). The first approach was defined by White (1982) and contains the porosity ε, the mass specific surface area S_m and the solid density ρ_s (Eq. 1.24). A second approach (Eq. 1.25) is traced to Kozeny (1927) and Carman (1956) assuming approximately spherical particles with the Sauter diameter d_{32} and a shape factor ψ. The Wadell shape factor ψ_{Wa} can be inserted for ψ:

$$r_p = \frac{2 \cdot \varepsilon}{(1 - \varepsilon) \cdot \rho_s \cdot S_m}, \qquad (1.24)$$

$$r_p = \frac{\psi \cdot d_{32} \cdot \varepsilon}{3 \cdot (1 - \varepsilon)}. \qquad (1.25)$$

Another basis for studying the wetting of powders is the Darcy equation (Eq. 1.26) which deals with a flow through a porous media (Darcy, 1856). l is the length, B is the permeability of the porous media and Δp is the driving pressure gradient:

$$\frac{dl}{dt} = -\frac{B \cdot \Delta p}{\eta \cdot l}. \qquad (1.26)$$

The liquid penetration through porous media was also studied by Kozeny (1927) and Carman (1956). They derived an equation (Eq. 1.27) by inserting the hydraulical pore radius (Eq. 1.25). j is a factor depending on the particle size, the particle shape and the porosity and S_v is the volume specific surface area:

$$\frac{dh}{dt} = -\frac{\Delta p \cdot \varepsilon^2}{j \cdot \eta \cdot h \cdot (1 - \varepsilon)^2 \cdot S_v}. \qquad (1.27)$$

Several authors studied the shape and orientation of pores in porous systems and tried to introduce them into their models, instead of working with equivalent pore radii (Benavente et al., 2002, Czachor, 2006, Staples and Shaffer, 2002). Two examples, the approach of Czachor (2006) and of Staples and Shaffer (2002) are shown in Figure 1.5. Benavente et al. (2002) introduced correction factors for the pore and the pore orientation which is discussed more detailed in subsection 1.3.1.

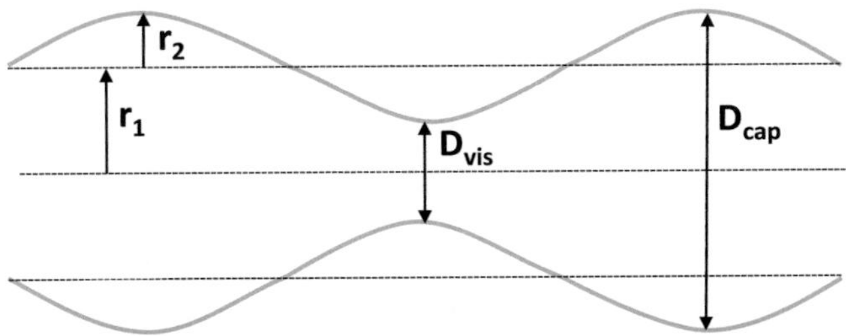

FIGURE 1.5: Schematic sketch of models for sinusoidal pores of Czachor (2006) and Staples and Shaffer (2002).

The approach of Staples and Shaffer (2002) is based on two characteristic diameters D_{cap} and D_{vis} which represent the capillary pressure and the viscous drag to express the real pores and it is also explained in subsection 1.3.1. Czachor (2006) modelled the capillary rise in axi-symmetrical sinusoidal pores which are described by two radii, the mean capillary radius r_1 and the waviness amplitude r_2, and the characteristic medium structure length h (Eq. 1.28). x gives the location in the capillary:

$$r(x) = r_1 + r_2 \cdot \sin\left(\frac{\pi \cdot x}{h}\right). \tag{1.28}$$

Czachor (2006) indicated that the capillary pressure is not constant in sinusoidal pores but has to be positive for spontaneous liquid penetration. Therefore, he introduced a critical contact angle θ_{crit} which decides about imbibition or not. Overall, he found a strong dependency of the waviness of sinusoidal pores on liquid penetration. By neglecting the pore structure and applying the Washburn approach to determine the contact angle of a powder, the values were between 70° and 85° for glass beads. The true contact angle was measured to be 27°.

1.2 Capillary wetting into food powders

Capillary wetting into food powders is a key factor for manufactures as well as for consumers. During production of particulate food material a poor wettability would probably lead to either an extended processing time or to low product quality (Charles-Williams et al., 2013). Powdered food products are usually reconsituted in a liquid by the consumer before usage. The reconstitution or rehydration process is divided into several steps under which the wetting is often the rate limiting one and, therefore, of major interest (Schubert, 1990).

1.2.1 Reconstitution of food powder

Drying of raw food material with the objective to manufacture powdered food gets more and more attractive for food companies with regard to the enhancement towards stability, storage, transport, weighing and processing (Cuq et al., 2011). Since food material includes a lot of different ingredients demonstrating different behavior during drying, the drying step towards the production of food powders has been studied by several authors (Bhandari et al., 1993, Chen and Patel, 2008, Reineccius, 2004). But for consumption very often a rehydration process is necessary in order to transfer a dry powder back into a liquid (Fang et al., 2007). During rehydration of a food powder generally four steps can be distinguished: 1. the wetting of the powder surface, 2. the sinking, 3. the dispersing and 4. the dissolution in case of soluble material (Pfalzer et al., 1973, Schubert, 1990). These steps take place consecutively but with overlapping and, thus, it is difficult to study them separately. The wetting is also described as wetting with following penetration of the liquid into the pore system due to capillary forces (Forny et al., 2011). A sketch of the four reconstitution steps is provided in Figure 1.6. Summarizing the four steps, they are well-known as instant properties. To declare food powders as instant powders, the whole reconstitution process has to be completed within a few seconds for a bulk height of 10 mm (Schubert, 1990).

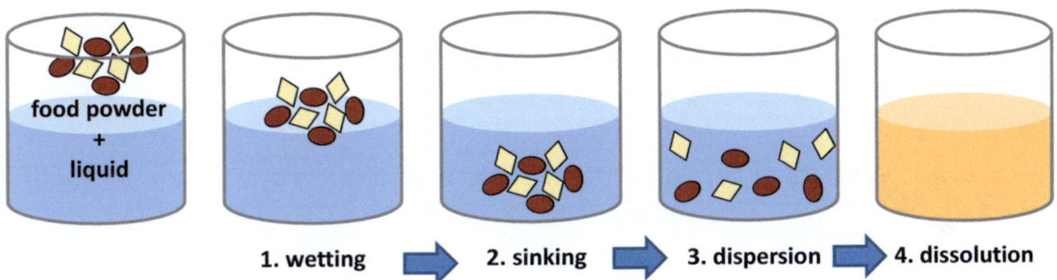

FIGURE 1.6: Four steps of food powder reconstitution: wetting, sinking, dispersion and dissolution.

Undesired phenomena can occur during reconstitution processes such as lump formation and sedimentation. Therefore, it is often necessary to modify food powders for enhancement of their instant properties. An important parameter which has to be controlled is the particle size (Freudig et al., 1999, Hogekamp and Schubert, 2003). Small particles (100 µm) build a narrow pore network in which the liquid penetrates. Due to dissolution a thick gel-like mass forms in these small pores that resists a further liquid penetration (Ortega-Rivas et al., 2006). On the other hand, if the particle size and, thus, the pore size is too large, the compensation of capillary pressure is reached before the pores are fully wetted by the liquid. Additionally, the dissolution time increases with increasing particle size. An average particle size of 400 µm was found to be an optimum (Freudig et al., 1999). The particle size of food powder can be adapted by granulation or agglomeration processes (Litster et al., 2004, Schubert, 1987). Besides the particle size, the porosity can also be a critical parameter (Schubert, 1978). It has to be considered that two porosities are important during rehydration. First, the liquid has to enter the large pore space around the particles, agglomerates or granules. Secondly, the fine pore network within an agglomerate or porous particle has to be filled with liquid (Schubert, 1990). The third important parameter which significantly influences the reconstitution is the contact angle between liquid and solid. Schubert (1990) stated that a food material showing a contact angle of $\leq 60°$ needs no modification in terms of instantization.

In case of higher contact angles, the surface properties can be changed by spraying of lecithin or by removal of surface fat, for instance (Schubert, 1987, 1990).

For improvement of instance properties of a food powder, the rate limiting step has to be identified and subsequently accelerated (Schubert, 1990). Mitchell et al. (2015) studied the whole reconstitution process of amorphous maltodextrin as model food powder and developed a mapping strategy in order to identify the limiting rate constants. Dupas et al. (2017) investigated the reconstitution of carbohydrate powders comparing crystalline and amorphous materials. They found dissolution being the rate limiting step during the reconstitution of large crystalline sucrose particles. The performance of amorphous maltodextrin during wetting, capillarity and dissolution is of high complexity depending on moisture content, molecular weight, particle size and water temperature.

1.2.2 Wetting step as rate limitation

As the first step, the wetting has a major role within rehydration, often being the time limiting factor (Schubert, 1990). Thus, many authors have focused on studying the wetting step during food powder reconstitution in terms of different impact factors and using different methods (Dupas et al., 2015, Forny et al., 2011, Galet et al., 2004, Roman-Gutierrez et al., 2003).

Dupas et al. (2015) used a static wetting test to characterize the wettability of food powders and proposed a model to predict the wetting time. Differently prepared soymilk powders were investiagted in terms of their wettability by Jinapong et al. (2008). They found an improved wettability after agglomeration of spray dried particles. In both studies, the powder was poured on a static liquid surface and the time until the powder sinks was determined.

Roman-Gutierrez et al. (2003) applied a different technique to characterize the wettability of wheat flour and its components. Sessile drop method was performed on compacts which were prepared by pressing. The contact angles were captured for time periods of up to 300 s in order to investigate additionally the water drop absorption. While the starches and the sugars exhibit low contact angles and high rates of water drop absorption, gluten shows the lowest wettability properties (high contact angle, low absorption rate). The wheat flour can be ranked in between the protein and the carbohydrates.

Another method to investigate the wettability of powders was applied by Forny et al. (2011). They measured penetration rates of liquid into single agglomerates and powder beds by Washburn technique with the aim to model an entire rehydration process.

The static and dynamic water wetting of cocoa powder which represents a highly hydrophobic food powder was investigated by Galet et al. (2004). Sessile drop method was applied as static test and Washburn method was used to study the dynamic behaviour. Additionally, they applied the Steven's method (Stevens et al., 1974) where powder is poured on a liquid with known surface energy. All data was used to estimate a surface energy of cocoa powder of $30\,\mathrm{mJ\,m^{-2}}$ which confirms the very poor wettability in water.

1.2.3 Heterogeneity due to geometry

Heterogeneity makes wetting complex whereby we distinguish between geometric and chemical heterogeneity. The geometric heterogeneity again can be separated into heterogeneity due to roughness, heterogeneity due to pore shape and orientation and heterogeneity due to particle shape. Figure 1.7 provides an overview about the different influencing factors of geometric heterogeneity on capillary rise. As already mentioned in subsection 1.1.1, the impact of surface roughness on the wetting expressed by the contact angle was studied and described by Wenzel (1936, 1949) and experimentally confirmed by several authors (Busscher et al., 1984, Palzer et al., 2001).

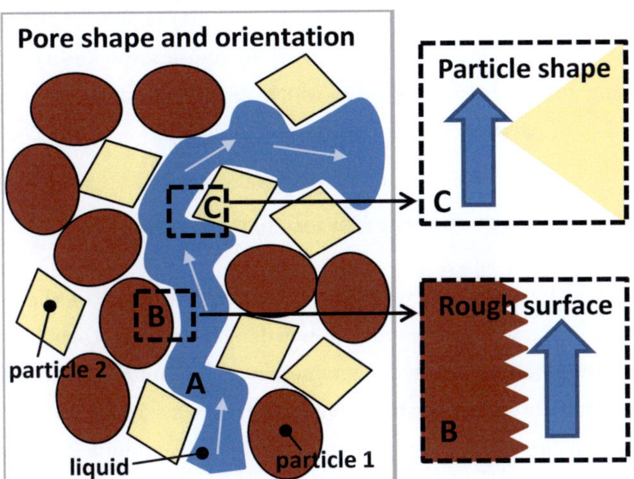

FIGURE 1.7: Geometric heterogeneities in pore networks within a powder bed influencing the capillary flow: A. variation in pore shape and orientation, B. roughness on particle surface, C. shape of particles.

The pore shape and orientation within powdered system is influenced by the size of the particles but also by the particle size distribution (Dang-Vu and Hupka, 2009, Kiesvaara and Yliruusi, 1993, Kirdponpattara et al., 2013). Dang-Vu and Hupka (2009) focused on the influence of particle size and particle size distribution on the capillary rise into glass powder. They used a small ($60\,\mu\mathrm{m}$ to $110\,\mu\mathrm{m}$) and a large fraction ($150\,\mu\mathrm{m}$ to $250\,\mu\mathrm{m}$)

of the powder and a mixture of both for their study. Small particles with a narrow distribution form small pores resulting in an even liquid flow, while the penetration into the large fraction results in an uneven behaviour due to diversified pore sizes. Mixing both fractions again improves the homogeneity of the bed since small particles fill the pore space between large particles. Kiesvaara and Yliruusi (1993) studied the effect of particle size on the wettability of lactose powder and found the fastest rise into the powder consisting of the largest particles ($> 212\,\mu m$). They explained the slower rates for the fine fraction ($< 106\,\mu m$) by a more effective degree of packing and, therefore, the formation of smaller pores. On the contrary, Kirdponpattara et al. (2013) determined decreasing penetration rates into nylon powder with increasing particle size ranging from $< 250\,\mu m$ to $500\,\mu m$ - $2000\,\mu m$. They explained this phenomenon with the presence of voids in the larger particle size fractions which act as wetting barrier. But they also observed increasingly heterogeneous flow profiles with increasing particle size as reported by Dang-Vu and Hupka (2009). By mixing all size fractions, the penetration rate increases again due to filling of voids with smaller particles and the flow profile gets more homogeneous.

Besides the particle size, the shape of the particles affects the powder packing and, hence, the capillary wetting. Yekeler et al. (2004) expressed the particle shape by four parameters, the roundness, the elongation, the flatness and the relative width. With increasing roundness and relative width and decreasing elongation and flatness, the wettability of particles was improved.

1.2.4 Heterogeneity due to surface composition

The chemical heterogeneity of food powders is related to their composition. Many food powders are multicomponent powders containing carbohydrates, proteins, fats, water and micronutrients such as minerals and vitamins (Bhandari et al., 2013). While carbohydrates are generally hydrophilic and, thus, easily wetted with water based liquids, hydrophobic fats repel water. Even if proteins are amphiphilic macromolecules they also tend to show water-repellent properties (Bhandari et al., 2013). Since wetting takes place on the surface of a solid the composition of the particle surface and not the bulk composition determines the interactions between liquid and solid (Gaiani et al., 2006). Therefore, Gaiani et al. (2006, 2010, 2007) studied the effect of surface composition of dairy powders on their wetting properties and the influence of drying and storage conditions on the surface composition. They found an increasing surface concentration of fat with decreasing drying temperatures in a spray dryer and a linear correlation of this surface concentration and the wetting time. The higher the surface concentration of fat, the higher the wetting time. An addition of lactose to the spray dryer feed significantly decreases the wetting time (Gaiani et al., 2010). Furthermore, an accumulation of fat at the particle surface could be observed for all tested dairy powders and a further concentration increase during storage of the powder samples. Thus, the wetting time also increases with storage time. Higher temperature and the absence of water tight storage bags accelerates the fat migration to the surface. The decreasing wetting properties with storage can not only be explained by fat migration, but also by particle collapse or caking of lactose depending on the glass transition (Gaiani et al., 2007). The impact of surface composition in terms of lactose, protein and fat on the wetting behavior of four different milk powders was also studied by Kim et al. (2002). The wettability of the powders was investigated before and after fat extraction. They found a strong decrease in wetting time after fat extraction for the two powders with initially high fat content, while the other two powders with high protein content did not show improved wetting behavior after fat extraction. Consequently, milk proteins as well as fat on the surface act as hydrophobic components slowing down the wetting step. Another study which deals with the wetting behaviour of dairy powders was performed by Crowley et al. (2015). They measured dynamic contact angles on tablets made of different milk protein concentrate powders. With increasing protein content in the powders and decreasing lactose content, the initial contact angles increases and the spreading rate decreases. According to their wetting properties, the powders were categorized into three groups, low-, intermediate- and high-protein powders.

The wettability of model cocoa drink powders, which consisted of coated or uncoated mixtures of cocoa with sugar, maltodextrin or milk powder was investigated by Kowalska and Lenart (2005). In the uncoated case, the fastest wetting times were achieved for the powders with the highest carbohydrate content (sugar or maltodextrin). Coating with hydrophobic components, such as cocoa or milk powder, increased the wetting time, while a sugar coating decreased the wetting time even if the bulk composition is mainly hydrophobic. Twelve commercial food powders were analyzed regarding composition, contact angle and wetting time in a study of Fitzpatrick et al. (2016). They observed a clear correlation between high surface content of carbohydrates combined with low fat and protein content and low contact angles and short wetting times. These three studies emphasize the relevance of knowledge about particle surface compositions and the interactions of each component with the wetting liquid.

Besides the composition of food particle surfaces, the internal structure of a component also influences the wettability. While the wetting performance of amorphous carbohydrates strongly depends on the water activity, this factor does not have to be considered for crystalline carbohydrates. Dupas et al. (2017) found that the contact angle of amorphous maltodextrin DE2 decreased from $30°$ to $15°$ with an increase of water activity from 0.15 to 0.73. The contact angle of crystalline sucrose was measured to be $16°$.

Kammerhofer et al. (2018a) studied the influence of an increasingly hydrophobic component in a single gap and in a powder mixture in order to emphasize the coupled influence of pore shape and hydrophobicity. They found a consequent influence of an increasingly hydrophobic contact angle on wettability in a single gap, while the wetting time of the powder mixtures is not influenced by this factor at low concentrations. Furthermore, a critical concentration of hydrophobic material was identified above which the wetting behaviour is dominated by the hydrophobic surface in the sample. This phenomenon was also observed by Nguyen et al. (2009) who measured drop penetration times into mixtures of hydrophilic lactose and hydrophobic salicylic acid powder. Depending on the particle size of the two powders this critical concentration was between 7 % and 17 % of hydrophobic mass.

1.2.5　Dissolution and swelling during wetting

As already mentioned above, the four reconstitution steps overlap and influence each other. Thus, during the wetting step of soluble food material dissolution of the wetted surface starts immediately. This can lead to an alteration of liquid and solid properties which again affects the wetting performance. An overview about effects of dissolution on liquid and solid properties is presented in Figure 1.8. In case of liquid properties, the wetting is mainly controlled by viscosity, density and surface tension. While dissolution of surface active components such as most proteins or lecithin significantly reduces the surface tension (Yang et al., 2008), dissolved biopolymers such as polysaccharids increase the viscosity and density of the wetting liquid (Wangler and Kohlus, 2017).

Schubert (1990) proposed the spraying of surface active components such as lecithin on food powders with insufficient instant properties such as full milk powder. During dissolution the surface active substance reduces the surface tension of water which improves the sinking. Mitchell et al. (2017) confirmed this surface tension effect by presenting a force balance for a single sphere on a liquid surface with the gravity force, the capillary force and the buoyant force. The capillary force presented in Eq. 1.29 shows that a reduction of surface tension, decreases the capillary force and, hence, favours the sinking step. z_c is the distance between the undisturbed liquid surface and the depressed liquid surface at the contact locus and ϕ is the angle between the horizontal and the solid-liquid-vapour contact locus. But they demonstrated a second surface tension effect counteracting the first one. Considering the Washburn equation (Eq. 1.15), it is obvious that a decreasing surface tension decreases the capillary rise into pores and, thus, the wetting performance:

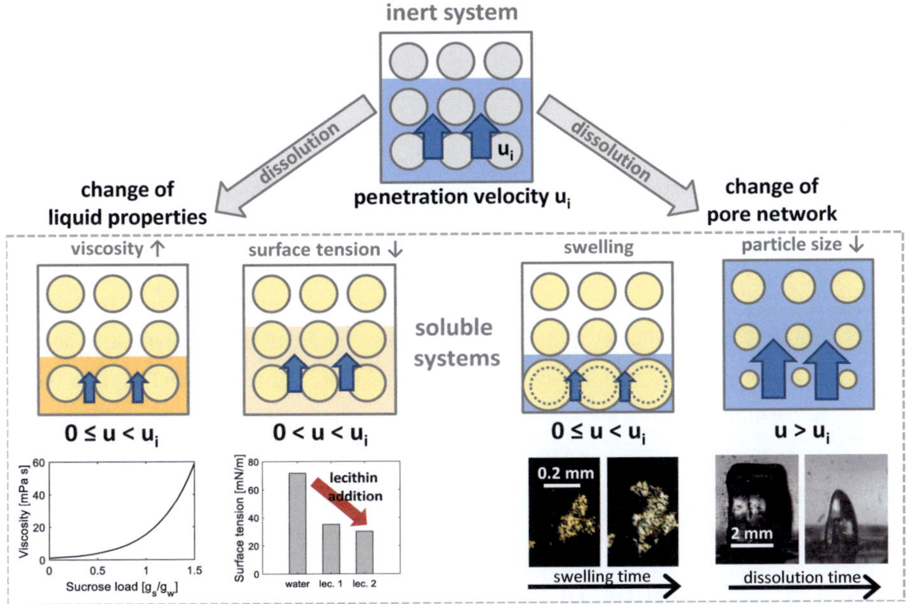

FIGURE 1.8: The effect of dissolution during capillary wetting on liquid properties and pore network.

$$F_c = \sigma \cdot 2 \cdot \pi \cdot r_p \cdot \sin\phi \cdot \frac{z_c}{\sqrt{1 + z_c^2}}. \tag{1.29}$$

The viscosity effect during wetting was recently studied by several authors (Dupas et al., 2017, Forny et al., 2011, Kammerhofer et al., 2018b, Wangler and Kohlus, 2017). Generally, it is known that viscosity development during wetting lowers the wetting performance. Forny et al. (2011) investigated the impact of increasing viscosity on wetting rates into agglomerates made of dextrose syrup. They found a linear relation of increasing viscosity and decreasing wetting rate. Wangler and Kohlus (2017) investigated the viscosity development during the wetting of biopolymers. Viscosity increases of up to 60 times within 200 s were observed. A viscosity increase of about 4 times within a very short time interval of 6 s during the wetting of sucrose was simulated by Kammerhofer et al. (2018b).

Besides the liquid properties, the pore structure of a powder bed is also influenced by dissolution. Since solid material is transported into the liquid phase, the particle mass and, thus, the particle size decreases. A reduced particle volume results in more free pore space for liquid uptake which favours the wettability. The dissolution rates for crystalline sucrose at three different temperatures were determined by Dupas et al. (2017). They measured values of $3.5\,\mu\mathrm{m\,s^{-1}}$, $7.8\,\mu\mathrm{m\,s^{-1}}$ and $44.5\,\mu\mathrm{m\,s^{-1}}$ for liquid temperatures of

5 °C, 22 °C and 80 °C, respectively. Slower dissolution rates were obtained for amorphous maltodextrin cuboids with a varying porosity at the same temperature. On the other hand, the increase of pore volume may also lead to a bed collapse resulting in the stop of the capillary flow (Kammerhofer et al., 2018b).

Another important factor which has to be considered during wetting of many food products is the swelling effect. Especially, protein and starch particles tend to increase their volume if they get in contact with water (Schubert, 1990). Due to this volume increase the pore volume decreases resulting in a reduced wetting performance. For simulations of capillary wetting into channels made of amorphous maltodextrin, Dupas et al. (2017) took swelling into account and set the maximum swelling to a factor of 5 %. In case of crystallin sugars, swelling does not have to be considered. For milk protein concentrate particles, an increase in particle size due to swelling from 300 µm to 600 µm was reported (Selomulya et al., 2013).

1.3 Modelling of the reconstitution process

A general aim in food industry is the introduction of reliable methods to predict the behavior of food powder during rehydration. This can be achieved by modeling the process, either with empirical models or with physically based models. Since a reconstitution process consists of several different steps, Forny et al. (2011) proposed the implementation of one model dealing with the entire reconstitution process where each step is represented by a set of physical equations. This approach is useful for the optimization of food powders and their rehydration performance. In literature, the wetting step is generally posed as capillary wetting process and, thus, driven by capillary forces (Cuq et al., 2011, Forny et al., 2011, Freudig et al., 1999, Hammes et al., 2018, Saguy et al., 2005b, Schubert, 1990). The Washburn equation (Washburn, 1921) and its modified versions offer opportunities for a physical description of the phenomenon. The ability of particles or agglomerates to immerse in water is determined by their density and particle size and the resulting weight force, while the buoyancy and the surface tension counteract (Raux et al., 2013). Hence, this step can be physically described by a force balance. A submerged agglomerate has to break into primary particles requiring a sufficient amount of energy to overcome the strength of the agglomerate (Forny et al., 2011). The tensile strength of solid bridges in agglomerates can be predicted by the Rumpf equation (Rumpf, 1958). Consequently, the dispersability of primary particles is often expressed by the sedimentation velocity which is also derived from a force balance (Hogekamp and Schubert, 2003). The most complex step in food powder reconstitution in terms of physical correlations is the dissolution. Approaches which are suitable to describe this step are the mass transfer equation (Fitzpatrick et al., 2016) or diffusion expressed by Fick's law (Wang et al., 2008). Hereinafter, an overview about models predicting the wetting step, the dissolution or both steps combined which are available and used in literature is presented.

1.3.1 Modelling the wetting step

If it comes to the modelling of the wettability of powders, it can be distinguished between different approaches. Most models focuse on wettability due to capillary forces which is called capillary wetting and are based on the Washburn equation (Washburn, 1921) (Eq.1.15). These models do not only deal with the contact between liquid and solid expressed by Young equation (Young, 1805) (Eq.1.3), but they include already the immersion of liquid into the pore network of a powder.

Saguy et al. (2005b) suggested the adoption of knowledge developed in other fields such as in soil science, but also highlighted the difficulties by applying capillary imbibition theories from other research areas to the water uptake of food materials. Established models can be used but have to be adapted to food phenomena such as dissolution or swelling. They applied an approach (Eq. 1.30) of Benavente et al. (2002) developed for capillary imbibition of porous rocks for modelling the rehydration of freeze-dried carrots (Saguy et al., 2005a). C_{theo} presents the theoretical penetration rate, ε the powder porosity, ρ_l the liquid density, r_p the pore radius, σ the surface tension of the liquid, θ the contact angle and η the liquid viscosity. Discrepancies between theoretical and experimental penetration rates was observed and explained by the deviation of the condition of pores in food materials from effective cylindrical capillaries. Pores in a porous medium are characterized by a pore size distribution, various shapes and geometric deviation from cylindrical. Furthermore, the model assumes a constant contact angle which is also not the case due to dynamic behaviour and heterogeneity in food stuff (Kammerhofer et al., 2018b). In addition, the pore system in carrots changes continuously during rehydration due to swelling:

$$C_{theo} = \varepsilon \cdot \rho_l \cdot \sqrt{\frac{r_p \cdot \sigma \cdot \cos\theta}{2 \cdot \eta}}. \tag{1.30}$$

As already mentioned above, the real shape and orientation of pores within a porous system is often neglected and leads to deviations from experimental data. Therefore, Benavente et al. (2002) implemented a correction factor for the pore shape, the roundness δ, and a correction factor for the pore orientation, the tortuosity τ. Moreover, they replaced the pore radius r_p by a characteristic hydraulic radius r_{eff} obtained by effective medium approximation (EMA). This radius does not only provide information about the pore size distribution, but also on the topology of the pore network. The alteration of Eq. (1.30) results in Eq. (1.31):

$$C_{theo} = \varepsilon \cdot \rho_l \cdot \sqrt{\frac{\delta \cdot r_{eff} \cdot \sigma \cdot \cos\theta}{2 \cdot \tau \cdot \eta}}. \tag{1.31}$$

Another approach which deals with the shape of the pores within porous systems was derived by Staples and Shaffer (2002). They developed a two diameter equation (Eq. 1.32) and found a good approximation for simulated flow curves in sinusoidal tubes. The model equation includes two characteristic diameters D_{cap} and D_{vis} which represent the capillary pressure and the viscous drag, respectively. L is the distance, L_{eq} is the equilibrium height and C is the ratio of D_{vis} and D_{cap}:

$$ln\left(1 - \frac{L}{L_{eq}}\right) + \frac{L}{L_{eq}} = -\frac{C^2 \cdot D_{cap}^2 \cdot \rho \cdot g}{32 \cdot \eta \cdot L_{eq}} \cdot t, \tag{1.32}$$

with

$$L_{eq} = \frac{4 \cdot \sigma \cdot \cos\theta}{D_{cap} \cdot \rho \cdot g}. \tag{1.33}$$

Fries and Dreyer (2008a) modified the Washburn approach by considering besides the capillary and viscous term also the gravity term. This extended version enables the calculation of the capillary rise for longer time periods. The equation is presented and explained in subsection 1.1.2 (Eq.1.16).

Martic et al. (2002) introduced the dynamic contact angle into the Washburn approach in order to model the liquid imbibition into porous systems resulting in Eq. (1.34). P_1, P_2 and P_3 are constant parameters containing liquid and solid properties and interactions between liquid and solid. They found that the contact angle depends on the wetting rate during the early stages of the wetting process leading to a reduction of penetration rate:

$$h(t) = -\frac{P_2}{P_3} + \sqrt{\frac{2t}{P_3} + \left(P_1 + \frac{P_2}{P_3}\right)^2}. \tag{1.34}$$

The time t_{wet} which is needed to wet a powder bed consisting of spherical particle was derived by Schubert (1990). The equation (Eq.1.44) bases on the Darcy and Carman-Kozeny equations (Carman, 1956, Darcy, 1856, Kozeny, 1927) for a flow through a porous medium and on the capillary pressure and is, therefore, also Washburn based. H is the height of the powder bed and d is the diameter of the particles:

$$t_{wet} = \frac{15 \cdot (1 - \varepsilon) \cdot \eta \cdot H^2}{\varepsilon \cdot d \cdot \sigma \cdot \cos\theta}. \tag{1.35}$$

Other models are based on the force balance between weight force, capillary force and buoyancy of a single sphere on a liquid surface (Dupas et al., 2015, Raux et al., 2013). These models combine the wettability and the sinkability. One model based on the force balance of floating spheres was introduced by Dupas et al. (2015) to obtain the percentage of powder crossing a static air-water interface within 45 s, which is a relevant timescale for food powder reconstitution. The wettability performance of commercial food powders was predicted by the model being in good agreement with experimental results. Raux et al. (2013) modeled the wicking of granular media at the surface of a liquid by determination of a critical contact angle θ^* below which spontaneous immersion occurs. It was calculated to be 51° for a monodisperse 3-dimensional compact pile of spheres where particles form a tetrahedral network. Experimentally measured critical contact angles were comparable, but slightly larger. A further deviation can be observed for diameters above 100 µm, since gravity which is neglected in the model gains importance for larger diameters. A shift of θ^* was modeled by introducing either polydispersity, a pressure gradient across the liquid-air interface or defects in the pile compacity.

The time a single droplet needs to penetrate into a loose powder bed can be expressed by the drop penetration time t_{dr} (Middleman, 1995). It is derived by the volumen flow rate driven by capillary rise into a bundle of capillaries representing the open pores at the interface between droplet and powder. V_0 represents the initial drop volume:

$$t_{dr} = 1.37 \cdot \frac{V_0^{2/3} \cdot \eta}{\varepsilon^2 \cdot r_p \cdot \sigma \cdot \cos\theta}. \tag{1.36}$$

Hapgood et al. (2002) extended this drop penetration model by incorporating macrovoids in loose packed powders. Macrovoids inhibit the liquid flow since they are not available as pathway. Therefore, the porosity is replaced by an effective porosity ε_{eff} in Eq. (1.36) which is calculated by Eq. (1.37). ε_{tap} is obtained by tapping the sample to get the closest packing:

$$\varepsilon_{eff} = \varepsilon_{tap} \cdot (1 - \varepsilon + \varepsilon_{tap}). \tag{1.37}$$

Another extension of this model was proposed by Nguyen et al. (2009) who included the effect of hydrophobicity. The presence of hydrophobic pores in a system results in a decrease of fractional open area to the droplet. The parameter ζ represents the proportion of the powder surface which is hydrophobic and inhibits the liquid penetration. The modified porosity ε^* can be seen below (Eq.1.38):

$$\varepsilon^* = \varepsilon \cdot (1 - \zeta). \tag{1.38}$$

The opposite mechanism of wetting is drying. Thus, both processes are strongly related and pore network properties play an important role. Tsotsas and Mujumdar (2008) developed pore network models which are powerful tools to study drying and the influence of the pore network structure. Pore network modelling showed that hydrophilic pores favoured the drying process while hydrophobic pores resulted in a delay (Chapuis and Prat, 2007).

1.3.2 Modelling of dissolution

Various models for dissolution of carbohydrates are available in literature (Hodges et al., 1993, Koiranen et al., 1999, Miller-Chou and Koenig, 2003, Parker et al., 2000, Peppas et al., 1994, Wang et al., 2002, 2008). Generally, it is important to distinguish between the dissolution of amorphous and crystalline material due to different behaviour and, thus, different modelling approaches.

Wang et al. (2002, 2008) gave an overview about models which were used in literature to model the dissolution of water-soluble polymers. They distinguished between diffusion based models as they were used for instance by Peppas et al. (1994), models based on the nth-order chemical kinetic equation and t_{50}-models which predict the time for the release of 50 % of a tablet. The chemical kinetic model showed the best agreement with their experimental data. Furthermore, they tested three different empirical models to predict the dissolution of guar gum powder. The logarithmic function gave a better fit than the first order kinetic model and the Weibull function.

A detailed summary about modelling the dissolution of amorphous polymers is presented by Miller-Chou and Koenig (2003). They listed five different main approaches and provided examples for each classification. *Phenomenological models with Fickian equations* are presented as models which try to physically describe dissolution by using Fickian conditions and considering moving boundaries in the system. Secondly, *models with external mass transfer as the controlling resistance to dissolution* were mentioned. A further model class are *stress relaxation models and molecular theories* which predict the polymer relaxation response to solvent uptake. *Transport models for swelling and scaling laws for chain disentanglement* calculate the polymer dissolution in the anomalous transport and scaling models. As a last resort, they present *continuum framework models* which deal with viscoelastic effects and mobility changes of the polymer during dissolution, while the solvent diffusion is described by anomalous transport models.

In the following, some models are presented more detailed in terms of characterization of the model, purpose and outcome. Each of them is adapted to a specific system.

Parker et al. (2000) developed an empirical model based on experimentally measured dissolution rates for dissolution of polysaccharide powders using a bimodal population of grains and lumps. They describe the lump formation as pore blocking condition due to the formation of highly viscous gel layers around the single grains. Applying the model it was determined that an optimum of grain size exists in order to avoid lump formation. Too small particles favours lump formation, while the dissolution is too slow for large particles.

The dissolution of lactose was modeled by Hodges et al. (1993). This model includes the mutarotation of α-lactose to β-lactose which is described by the first-order kinetics and the mass transfer from the crystal to the liquid which is expressed by the mass transfer equation. Furthermore, a lactose balance was used to match the rate of the decreasing particle size with the mass transfer rate in order to consider particle diminution. The model gives a good prediction for the dissolution of lactose crystals in the regions above the α-lactose solubility limit when comparing with experimental data.

Koiranen et al. (1999) used a population balance based dynamic model for the prediction of dissolution of sucrose crystals where the dissolution is assumed to be controlled only by the mass transfer. The aim of this study was the fitting of calculated population densities with experimentally determined mass of crystals.

A kinetic model for the dissolution of crystalline glucose and fructose in 2-methyl 2-butanol was established by Engasser et al. (2008) where they combined sugar dissolution and mutarotation kinetics. The inital dissolution step was assumed to be limited by the solute transport from the surface to the bulk solution. Therefore, they calculate a sugar transport coefficent by diffusion and convection using a mass transfer correlation for stirred tank reactors (Eq. 1.39). For small particles, they determined the Reynolds number Re from the Archimedes number Ar (Eq. 1.40):

$$Sh = 2 + 0.5 \cdot Re^{0.5} \cdot Sc^{0.33}, \tag{1.39}$$

$$Re = \left(\left(14.4 + 1.8 \cdot Ar^{0.5} \right)^{0.5} - 3.8 \right)^2. \tag{1.40}$$

1.3.3 Modelling of dissolution during capillary wetting

Recently, some authors have focused on modelling the coupling of the liquid penetration into a particle bed with the dissolution and the resulting effects such as the change of liquid properties and the change of the particle bed (Dupas et al., 2017, Hammes et al., 2018, Wangler and Kohlus, 2017). For all models, the capillary pressure is the driving force and the dissolution phenomenon is modeled by empirical correlations based on experimental investigations.

Dupas et al. (2017) developed an approach to simulate the capillary penetration in soluble channels of amorphous maltodextrin. The dissolution is described by an empirical equation depending on the liquid concentration, the moisture and the speed of the water penetration. The flow Q through the capillary is expressed by Eq. (1.41) where r_0 is the initial radius of the channel and z the direction of the flow. Furthermore, they implemented a simplified swelling model. Applying the model a viscosity increase due to an increased maltodextrin concentration in the channel and a resulting penetration stop can be observed. This effect strongly depends on the initial pore size and wall thickness. With increasing pore radius the covered distance increases:

$$Q(t) = \frac{\pi \cdot \sigma \cdot \cos\theta}{4 \cdot r_0 \cdot \int_0^L \frac{\eta(z)}{r(z,t)^4} dz}. \tag{1.41}$$

A modified Washburn model was presented by Hammes et al. (2018) who introduced a method to determine contact angles of soluble, particulate materials. The Dissolution-Modified Washburn model assumes much lower kinetics of the dissolution process compared to the capillary rise and slow dissolution leads to a linear increase of the inner capillary radius with time. They found an improvement in modelling the capillary rise process when partial dissolution of the bed in the wetting liquid takes place compared to the classical Washburn model. Estimates of contact angles which are independent of the sugar concentration of the liquid can be obtained with the model. They stated that the model is able to describe the two phenomena, wetting and dissolution, independently. The model equation is presented in Eq.(1.42) where k_{diss} is the dissolution rate. In Eq.(1.43), the correlation for the parameter C_{th} is given:

$$\frac{d^2h}{dt^2} + \left(\frac{8 \cdot \eta \cdot h}{\rho \cdot C_{th}} + \frac{1}{C_{th}} \frac{\partial C_{th}}{\partial t} \right) \cdot \frac{dh}{dt} - \frac{2 \cdot \sigma \cdot \cos\theta}{\rho \cdot r_0 \cdot h} + g = 0, \tag{1.42}$$

with

$$C_{th} = \int_0^h \left(r_0 + \frac{h-z}{h} \cdot k_{diss} \cdot t \right)^2 dz. \tag{1.43}$$

Wangler and Kohlus (2017) established a model dealing with the wetting behaviour of a dynamically changing powder system. The Washburn equation provides the basis of the model. It is modified by the addition of temporally and locally changing parameters due to dissolution and swelling. They consider the change of porosity, particle diameter and viscosity as it can be seen in Eq.(1.44). The porosity decrease was correlated to the volume increase due to swelling, while the total bed volume stayed constant. The dissolution rate which is the increasing concentration of polymer in the rising liquid was determined from viscosity-development measurements. Their model allowed a qualitative simulation of the capillary water uptake into a biopolymer powder depending on the influence of the changing parameter:

$$h = \sqrt{\frac{\varepsilon(t,x) \cdot d_{32}(t,x) \cdot \sigma \cdot \cos\theta_{eff}}{15 \cdot \eta(t,x) \cdot (1 - \varepsilon(t,x))}} \cdot t. \tag{1.44}$$

1.4 Objectives based on state of the art

The main objective of this thesis is the prediction of capillary wetting of heterogeneous and soluble food powders. In order to achieve this objective a detailed experimental study was peformed and a new model was developed. It is necessary to identify the main influencing factors on capillary wetting in order to consider those during model development. Therefore, the achievement of the main goal is realized by studying single influencing factors separately and combining them stepwise to enhance complexity.

Effect of heterogeneity: In a first step the effect of heterogeneity is investigated by performing capillary rise experiments with hydrophilic and hydrophobic walls in a single gap and with mixtures of stochastical distributed hydrophilic and hydrophobic particles on a powder level. The hydrophilic and hydrophobic materials are inert towards the contact with the wetting liquid implying no dissolution or swelling. An available model from literature is adapted and tested for heterogeneous, inert powder systems.

Effect of dissolution: Secondly, the effect of dissolution during capillary wetting is separately investigated by performing experiments with three hydrophilic and soluble food powders. For this purpose, a new model which is based on physical equations is developed predicting capillary penetration into soluble powders. The new model is validated with experiments.

Combination of heterogeneity and dissolution: Furthermore, the effect of heterogeneity and the effect of dissolution are combined in the experimental investigations by mixing hydrophilic, soluble and hydrophobic, inert particles. The new model is extended towards the effect of heterogeneity in terms of contact angle.

Application: In a last step, the developed model is applied for a real soluble food powder exhibiting heterogeneity in terms of composition on the surface and, thus, in terms of contact angle.

2

Experimental methods

In this chapter, all experimental techniques which were applied in the course of this thesis are presented and described. Powder properties were determined in order to characterize the particulate materials. Surface modification techniques were used to create defined attributes and surface analysis methods were performed for quantification. Furthermore, contact angles were measured by sessile drop method and setups for the investigation of liquid penetration into single gaps and powder beds described.

2.1 Analysis of powder properties

Different techniques were applied in order to characterize the powder properties of all particulate systems used in this thesis. The detailed specification of powders in terms of size and density is necessary for interpretation of wetting experiments. The following measurements were performed: determination of particle size distribution, analysis of solid density and characterization of bulk density and porosity. The particle size distribution of powders was determined by dynamic image analysis with a two camera system (CAMSIZER X2, Retsch Technology, Germany). Depending on the size and cohesivness of the particles either the free falling tool or the dispersion tool was inserted into the aparatus. The material density was measured by means of a helium pycnometer (Multivolume Pycnometer 1305, Micrometrics, USA) which works with a gas displacement method. Furthermore, the bulk properties were analyzed which are important for the study of the pore network within a powder. Therefore, the bulk density and the porosity were determined by weighing a powder sample in a defined volume after tapping the sample for 10 times.

2.2 Surface modification techniques and analysis methods

In this section, surface analysis and surface modification methods are presented. Those investigations are required to adjust and to quantify the surface properties of the materials which are used for penetration experiments.

2.2.1 Cleaning of glass material

Even if glass material is already of hydrophilic nature, a cleaning step is necessary to remove impurities from the surface leading to an increased hydrophilicity. Thus, the cleaning step is helpful to get reproducible surface conditions which are also used as pretreatment for the silanization prodecure. Hydrophilic glass material was obtained by applying a treatment with peroxymonosulfuric acid which consists of sulfuric acid and hydrogen peroxide in a volume ratio of 3:1, respectively, for 15 minutes. Afterwards, the glass was rinsed with distilled water until the waste water showed a pH-value of 7 and finally dried at 140 °C for 2 hours in an oven.

2.2.2 Silanization of glass material

Hydrophobic glass material is obtained by modification of the cleaned surface with a silanization process. The glass material was treated with triethoxyoctylsilane in toluene with different concentrations and for different times in order to achieve differently hydrophobic glass surfaces. The silanization parameters are summarized in table 2.1. The silanization treatments for 48 h and 72 h were performed with pre-dried toluene in order to reduce the water amount during the reaction. After silanization treatment the material was rinsed with toluene and acetone and dried under the fume hood.

TABLE 2.1: Silanization parameters for modification of glass material.

Hydrophobic material	Silanization concentration	Silanization time
A	10 mM	30 min
B	10 mM	48 h
C	20 mM	72 h

2.2.3 Fluidized bed coating

A coating process was applied in order to create a hydrophobic model food powder consisting of spherical glass beads with a thin layer of shellac. Further details about the used materials are provided in chapter 3. The fluidized bed technology was selected for the coating process. A scheme of the process is shown in Figure 2.1. Therefore, a 10 wt.% solution of Shellack in ethanol was atomized into the process chamber of a ProCell5 plant (Glatt, Germany) where 1 kg of glass powder was fluidized with an air flow of $43\,\mathrm{m^3\,h^{-1}}$. The spray rate was kept at 10 g/min with a pressure of 2 bar. For visualization of the coating quality, a blue dye was added to the spraying solution. A homogeneous colouring of the particle was taken as feature for a successfull coating process. In total, a mass of 334 g of coating solution was sprayed on the glass powders.

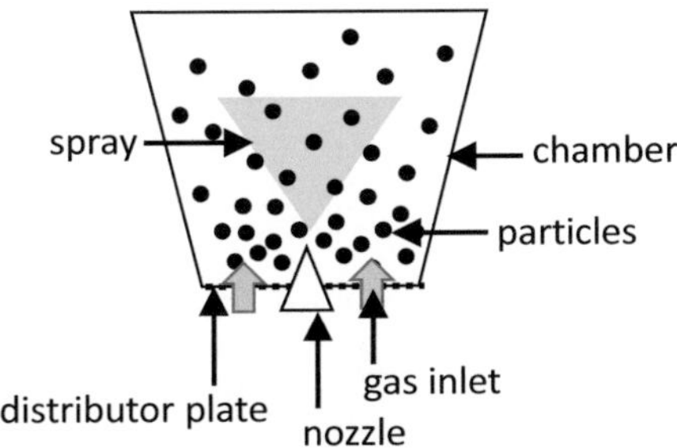

FIGURE 2.1: Scheme of coating process in fluidized bed.

2.2.4 Spin coating

The spin coating process was applied for the formation of thin layers of food materials on glass slides which are used for contact angle determination. A cleaned glass slide (see 2.2.1) was fixed by means of a vacuum pump in the spin coating device (SPIN 150, SPS-Europe, United Kingdom). Six droplets of the food solution were placed on top of the slide. The coating was carried out at 1000 rpm for 60 s. During this period the solvent completely evaporates from the surface and a thin layer forms. A scheme of the process is shown in Figure 2.2.

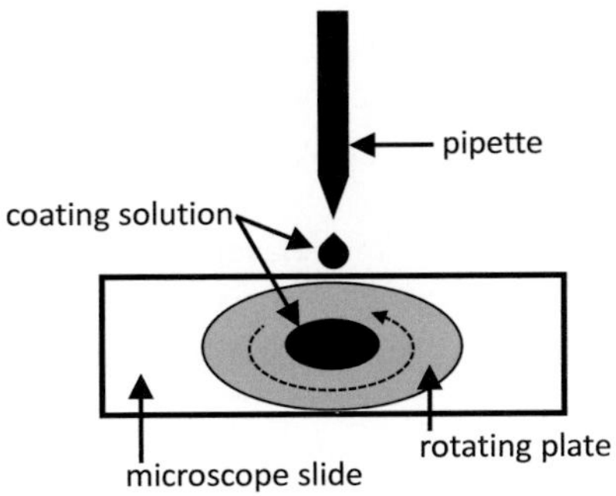

FIGURE 2.2: Scheme of spin coating.

2.2.5 Tablet pressing

The characterization in terms of contact angle of particulate food systems is challenging. One option to realize this task is the production of solid surfaces by pressing tablets and perform sessile drop method on those surfaces. Therefore, tablets of food powders with a diameter of 40 mm were compacted in a press (PWV 200 E, Paul-Otto Weber, Germany). The pressing was performed at 500 kN for 30 s which resulted in a pressing pressure of 398 MPa. A prepressing step was applied at 250 kN for 10 s (199 MPa). The high pressing pressure is required in order to create a surface with low porosity in avoidance of an immediate imbibition of the drop on the tablet. A scheme of the pressing process is presented in Figure 2.3.

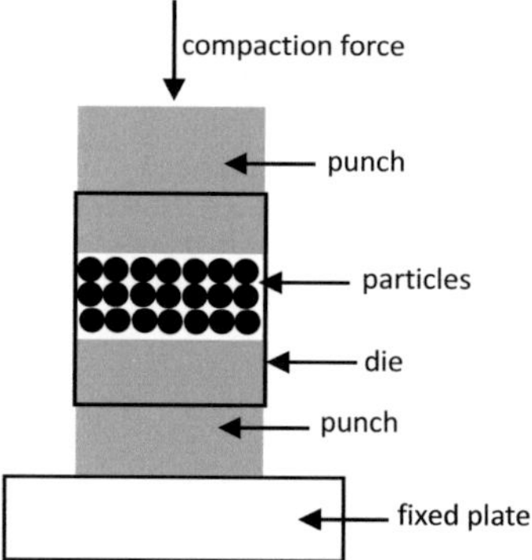

FIGURE 2.3: Scheme of tablet pressing.

2.2.6 Manufacturing of thin layers

A pressing step for the formation of flat surfaces for contact angle measurements is not applicable for food powders with a heterogeneous composition, since it would result in the alteration of the initial particle surface. Thus, in case of milk powder, thin layers of particles were created by means of an adhesive tape and later used for contact angle determination. A scheme of a monolayer on a adhesive tape is shown in Figure 2.4.

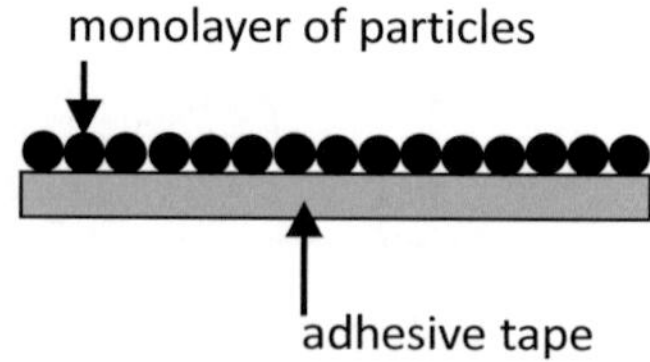

FIGURE 2.4: Scheme of monolayer of particles on adhesive tape.

2.2.7 Particle shape and surface roughness

Surface roughness was investigated qualitatively and quantitatively. Since it is well-known that glass material usually provides a smooth surface (Busscher et al., 1984, Palzer et al., 2001), the appearance of the surface was only qualitatively checked by means of a scanning electron microscope (SUPRA 55VP, Carl Zeiss, Germany). On the other hand, surface roughness of food tablets was quantitatively measured with focus variation with a light microscope using the InfiniteFocus device (Alicona, Austria). This measurement is necessary to get information about the surface condition of the tablet since surface roughness influences the apparent contact angle. The analysis was performed on three tablets and on each three times.

2.3 Contact angle determination

The water affinity of the different materials was measured by contact angle determination. The sessile drop method was performed on hydrophilic and hydrophobic glass slides, on food tablets and on thin layers of milk powder particles. A water droplet of $3\,\mu L$ was placed on top of the glass slide or tablet and a light source illuminated the setup from the back. A set of images over time was captured by a high speed camera (NX-S2, Imaging Solutions GmbH, Germany) equipped with a microscope objective from OPTEM ZOOM 125 until the static state was reached. Afterwards, the software ImageJ was used to determine the contact angles by analyzing the shape of the droplets. During the dynamic spreading of the droplet, additionally, the contact line speed was analyzed by measuring the droplet diameter over time. In Figure 2.5, the measurement of contact angle and droplet diameter can be observed. At least three droplets were used for contact angle determination.

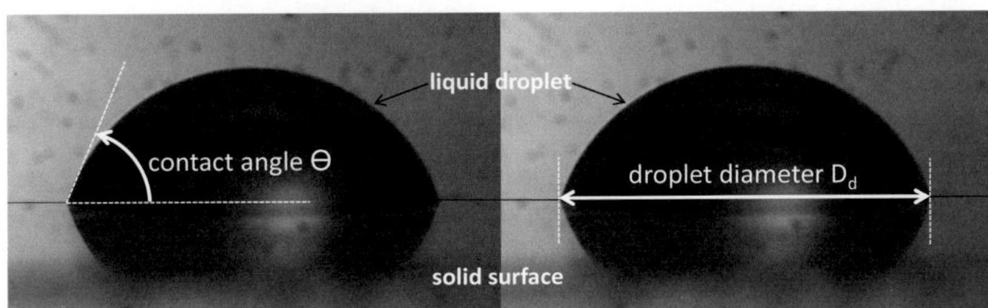

FIGURE 2.5: Analysis of contact angle and droplet diameter by means of recorded images.

2.4 Single gap setup

Single gap experiments were performed in a tailor-made setup which is shown in Figure 2.6 where two glass slides are fixed around a spacer. The spacer is changeable and determines the gap width. A movable beaker providing the water is installed below the gap setup and is moved upwards for starting the experiment. The water rise into the gap is captured by the same camera setup as used in the sessile drop experiment. For analysis of the images in terms of the equilibrium height the open-source software ImageJ is used.

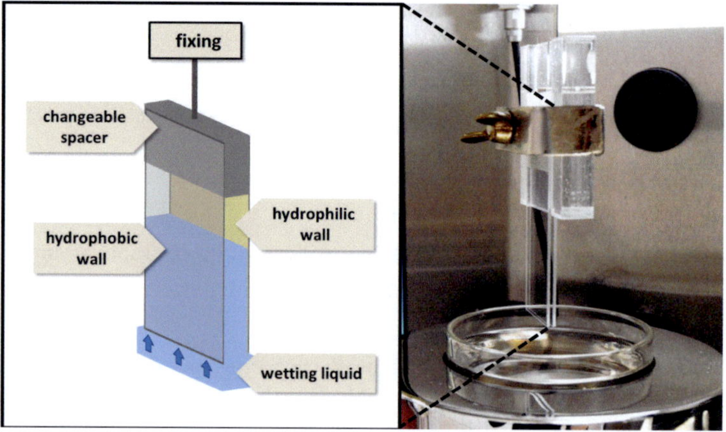

FIGURE 2.6: Experimental setup for liquid penetration into a single gap with adjustable gap width.

2.5 Washburn experiment

Experimental penetration rates for liquid into the different powders were measured by using the Washburn setup of the K100 tensiometer (Krüss, Germany). This equipment includes a cylindrical glass tube (radius r_C of 5 mm) which is connected to a balance. Below the tube a moveable and tempered container is installed wherein the wetting liquid is stored. A certain amount of powder (depending on the bulk density of powder) is filled into the cylinder and tapped 10 times in order to achieve a uniform packing. The bottom of the tube is closed with a filter paper. The penetration of liquid into the powder bed starts when the liquid gets in contact with the cylinder. The experimental setup can be seen in Figure 2.7.

The mass gain as a function of time is recorded as well as the liquid temperature. The measurement is controlled with LabDesk (Krüss, Germany). For our experimental

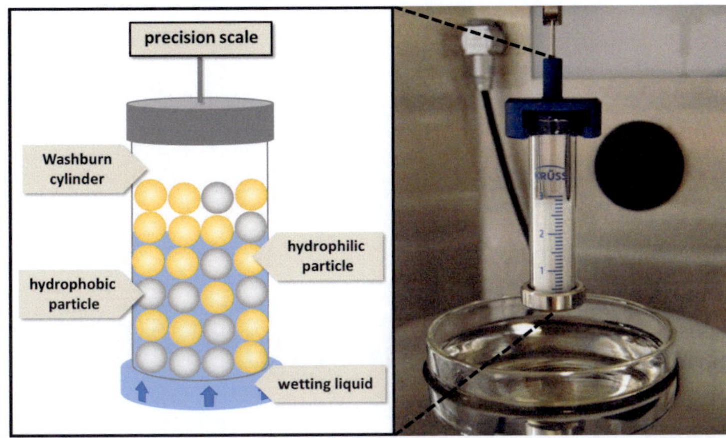

FIGURE 2.7: Experimental setup for determination of liquid penetration into powders (Washburn setup of K100 tensiometer, Krüss).

study, different mixtures of inert and food powders were tested. Water was used as wetting liquid. For each powder mixture, the measurement was repeated at least three times. All penetration experiments were performed at a water temperature of 20 °C in order to keep liquid properties such as density, viscosity and surface tension constant. Additionally, the penetration experiments were repeated with hexane as wetting liquid to determine the capillary constant K_{hex} of the powder packings.

3

Results of material characterization

This chapter provides the characterization of different inert and food powders which were used for wettability experiments in terms of composition, powder properties and solid liquid interactions. The detailed specification of the parameters is needed for interpretation and understanding of the wettability results, as well as for modelling and predicting the capillary wetting into inert, food or mixed systems. Glass material was chosen as inert model components since it does not react when getting in touch with wetting liquid. Hydrophilic and hydrophobic glass powders allow to investigate individually the effect of solid liquid interactions at the interface. Solubility and consequent effects such as viscosity increase were studied using homogeneous food powders such as sucrose, lactose and sodium chloride. In order to overlap single influencing factors, mixtures of powders were produced and investigated. Finally, wettability studies were performed with heterogeneous food powders containing hydrophilic and hydrophobic as well as soluble components which results in a highly complex system.

3.1 Composition of powders

Two soda lime glass powders (SiLibeads) were provided by Sigmund Lindner GmbH (Germany) with a content of 72% SiO_2, 13% Na_2O, 9% CaO and 4% MgO. The hydrophilic glass beads were obtained by a cleaning step as explained in subsection 2.2.1, while the hydrophobic glass beads were obtained by surface modification using a silanization step (subsection 2.2.2). Glass material is inert and, thus, does not react with the liquids used in this study.

Crystalline sucrose powder (Unser Feinster) was purchased from Nordzucker AG (Germany) and the desired size fraction was obtained by sieving. Sucrose is a disaccharide consisting of glucose and fructose. It is well soluble in water with a saturation solubility of about 200 g in 100 g water at 20 °C (Schwedt, 2010). Lactose powder SacheLac80 was provided by MEGGLE GmbH and Co. KG. It consists of α-lactose monohydrate. Lactose is a disaccaride made of glucose and galactose which has with 18.0 g in 100 g water at 20 °C a much lower solubility compared to saccharose (McSweeney and Fox, 2009).

Sodium chloride salt (Ursalz fein, SALDORO, Germany) was sieved into three different size fractions s_1, s_2 and s_3 and was also used as hydrophilic food component. The solubility of sodium chloride in water is with 35.8 g per 100 g at 20 °C in between the two disaccharides (Sawamura S. et al., 2007).

Besides the homogeneous powders consisting of only one component, also two milk powders which are heterogeneous in their composition were investigated in terms of wetting performance. A standard whole milk powder (WMP) and a whole milk powder with 80 % of free surface fat (WMP ff) were used to increase the complexity and study the influence of heterogeneous surface composition on wetting. WMP was provided by Nestlé Research Center (Switzerland) and WMP ff was purchased from Hochdorf Swiss Milk AG (Switzerland). The two whole milk powders are characterized by the same composition. They consist of 25 % protein, 41 % lactose and 27 % fat. The difference between WMP and WMP ff is the location of the fat in the milk particles. Due to a proper homogenization, the milk fat is equally distributed in the WMP. A missing homogenization step results in the accumulation of fat on the surface of WMP ff.

3.2   Shape and surface of particles

For characterization of the different powders in terms of their shape and their surface condition, a scanning electron microscope (SEM) was used to achieve an optical magnification of the particles. The method is described in subsection 2.2.7. For each powder, a magnification of 150 times and 650 times was applied which can be assigned to the images (a) and (b) in the Figures, respectively.

In Figure 3.1, the SEM-images of the coarse glass powder (a and b) and the coarse glass powder coated with shellac (c and d) are presented. Analyzing the images (a) and (c), it is obvious that glass powder and coated glass powder consist of very spherical and uniform particles with only a few exceptions. In the case of shellac coated glass powder some agglomerates can be identified as agglomeration can occure during coating. But in general, the shape of the glass beads did not change in the coating process. Zooming further in, the surface of the particles can be analyzed in the images (b) and (d). It can be seen that the particle surface of glass beads (b) is homogeneously smooth without any visible roughness. Even images with a magnification of 2000 (not shown here) did not show any roughness. In case of the coated glass beads, some irregularities in form of shellac droplets can be observed on the surface. Nevertheless, the overall surface smoothness remains unchanged after coating. Thus, it can be concluded that roughness is negligible for glass beads and coated glass beads.

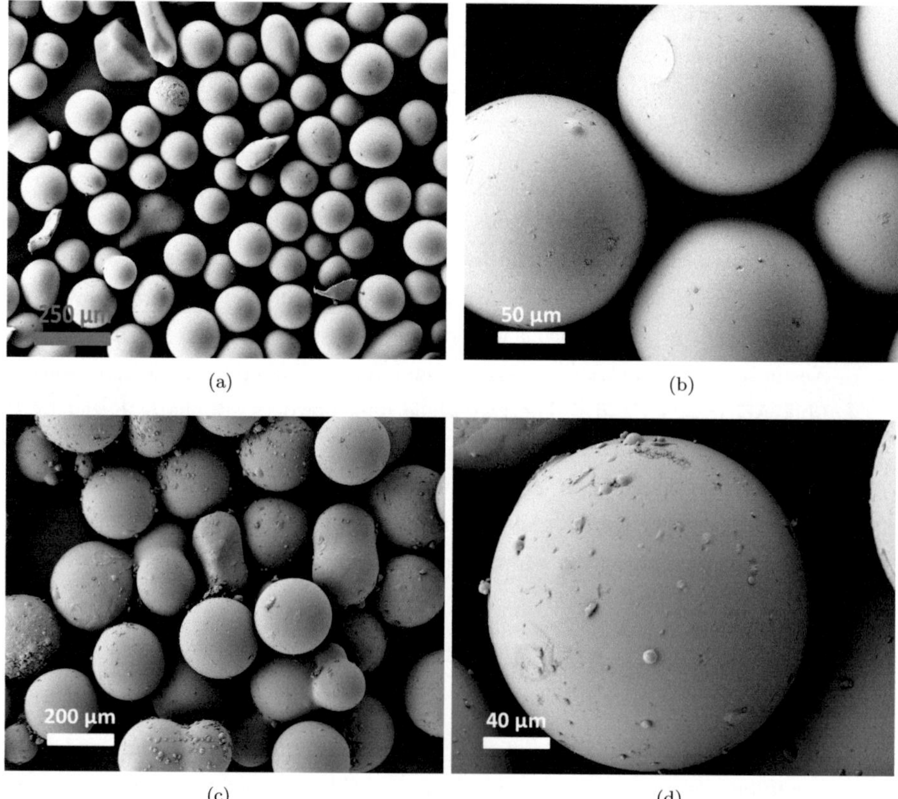

FIGURE 3.1: SEM-images of coarse glass beads with a magnification of 150 (a) and 500 (b) and SEM-images of shellac coated coarse glass beads with a magnification of 150 (c) and 650 (d).

Crystalline food powders such as sucrose, lactose and sodium chloride appear more heterogeneous compared to the glass beads. Figure 3.2 provides SEM-images of the three food powders with a magnification of 150 (a, c, e) and 650 (b, d, f). In a first step, the shape of particles is investigated (left images). Sucrose and sodium consists of angular particles with a very uniform size. On the other hand, the lactose particles are also of angular shape, but with a more heterogeneous size distribution. However, the shape of the three crystalline food powders is comparable with each other. Analyzing the surface condition of the three materials, in all cases the surface is covered by very fine particles. It seems that dust formed by attrition accumulates on the particle surface. Therefore, the particle surface of the three food powder appears heterogeneous with a certain roughness.

Figure 3.2: SEM-images of sucrose, lactose and sodium chloride with a magnification of 150 (a), (c), (e) and 650 (b), (d), (f), respectively.

Besides homogeneous, crystalline food powders, two heterogeneously composed food powders were analyzed by scanning electron microscopy. In Figure 3.3, the resulting

SEM-images are shown for WMP and WMP ff, each with a maginification of 150 (a+c) and 650 (b+d), respectively. The appearance is completely different to the homogeneous food powders. The particles of the milk powder consists of agglomerates with a very heterogeneous shape. Neither the size of the agglomerates, nor the shape shows a uniform condition. For WMP ff, the primary particles can be identified and, thus, the shape appears slighty more uniform.

FIGURE 3.3: SEM-images of WMP and WMP ff with a magnification of 150 (a+c) and 650 (b+d), respectively.

The surface of both milk powders appears smoother than the surface of sucrose, lactose and sodium chloride. Only some small droplets on the surface of WMP form some irregularities. The surface of WMP ff is mainly covered by dark areas, while the surface of WMP is homogeneously light. The dark areas consists very likely of surface fat, since 80 % of the WMP ff surface is covered by fat.

3.3 Characterization of powder properties

The powder properties are described by the particle size distribution, the porosity, the capillary constant and the apparent density. Characteristic size values such as the $d_{50,3}$-value or the specific surface area S_v are determined for specification of the particle size of the powders. The porosity and the capillary constant give the information about the free volume and the accessibility of pores for liquid penetration. Furthermore, the apparent density of the different powders is measured.

3.3.1 Powder properties of inert systems

Glass powder of two different size fractions (fine and coarse) was used as inert model material. Since it does not react with the wetting liquid the wetting behaviour can be investigated as single effect without consideration of dissolution or swelling. The material density of the glass was determined with a helium pycnometer which results in a value of $2469\,\mathrm{kg\,m^{-3}}$. Figure 3.4 provides the cumulative, volume based particle size distribution of both powders in dependance of the particle diameter. The particle diameter of the fine powder is in a range of about $70\,\mu\mathrm{m}$ to $110\,\mu\mathrm{m}$, whereas the coarse fraction has diameters of about $240\,\mu\mathrm{m}$ to $320\,\mu\mathrm{m}$. Additionally, the particle size distribution of coarse glass beads coated with shellac is shown in Figure 3.4 which is discussed at the end of this subsection.

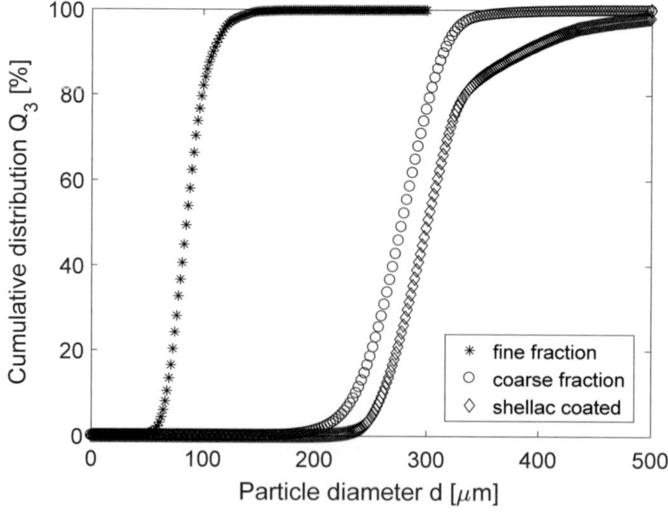

FIGURE 3.4: Cumulative particle size distribution Q_3 of fine fraction and coarse glass bead fractions and of shellac coated coarse fraction.

A summary of all characteristic values which are needed in this thesis are given in Table 3.1. $d_{10,3}$, $d_{50,3}$ and $d_{90,3}$ are characteristic particle diameters where 10%, 50% and 90 % of the mass or volume of the particles are below this value, respectively. Furthermore, the Sauter diameter d_{32}, the volume specific surface S_v and the sphericity Ψ are presented. The Sauter diameter, the sphericity and the resulting volume specific area are required to determine mixing ratios of inert and food powders. By comparing both size fractions, it is obvious that the volume specific surface area is almost four times higher in the fine fraction while the sphericity does not differ significantly.

TABLE 3.1: Characteristic size values of glass powders and glass powder coated with shellac.

Size values	Fine fraction	Coarse fraction	Shellac coated
$d_{10,3}$ [µm]	69.7	239.9	261.9
$d_{50,3}$ [µm]	86.9	280.7	301.5
$d_{90,3}$ [µm]	109.8	317.7	384.1
d_{32} [µm]	80.2	278.4	297.2
S_v [1/mm]	81.0	23.4	22.3
Ψ [−]	0.92	0.92	0.90

A further characterization of the powders is performed by quantifying the pore network. Thus, the capillary constants K_{hex} are determined using hexane as wetting liquid in the Washburn experiment which is described in section 2.5. Figure 3.5 (a) presents the capillary constants for the fine and the coarse glass powder. The pore geometry of the coarse fraction is expressed by a K_{hex}-value of $1.8 \, \text{mm}^5$ while the value for the fine fraction is smaller with $1.4 \, \text{mm}^5$. A higher capillary constant stands for a network which favours a better hexane penetration into the system. Additionally, the tapping porosities ε were measured and are presented in Figure 3.5 (b). The porosities of the fine and the coarse fraction with values of 0.39 and 0.40, respectively, do not differ greatly. Combining the information of Figure 3.5 (a) and (b), it can be concluded that both fractions create a dense powder bed due to high sphericity, but larger particles lead to larger pores and, thus, to a higher capillary constant.

Furthermore, the capillary constants and porosities are used to calculate effective pore radii with Eq. 4.5. The fine and the coarse powder exhibit effective pore radii of $2.7 \, \mu\text{m}$ and $3.9 \, \mu\text{m}$. The values do not stand for the real pore radii within the powder system, but express a substitute radius for a network consisting of a bundle of parallel pores, which would lead to the same hexane penetration.

Coarse glass particles were coated with shellac as described in subsection 2.2.3 in order to modify the glass surface. The coating leads to an increase of the particle size which can be seen in Figure 3.4. An averaged thickness of the coating layer was determined by the difference in $d_{50,3}$ of $20.8 \, \mu\text{m}$ before and after coating. Thus, the theoretical layer

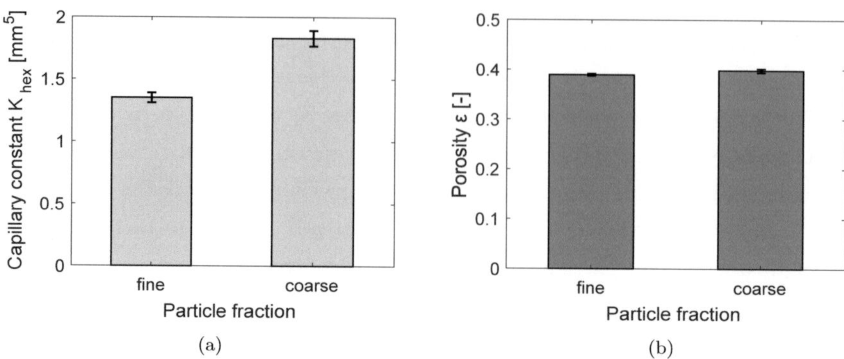

FIGURE 3.5: Capillary constant (a) and porosity (b) of fine and coarse glass powder fractions.

thickness was calculated to be 10.4 µm. This value can be solely seen as roughly estimated coating thickness since agglomeration and uneven distribution of coating material on the glass surface have an influence on coating layer.

3.3.2 Powder properties of food systems

The powder properties for the different homogeneous and heterogeneous food powders are provided in this subsection. For each powder, the particle size distribution with characteristic size values, the capillary constant, the porosity and the apparent density are presented.

3.3.2.1 Homogeneous food powders

Three homogeneous food powders were used in this study. Sucrose and lactose are representatives of the carbohydrate group, while sodium chloride is a mineral. The three powders are hydrophilic and water soluble and are contained in many powdered food products such as cocoa beverage powder, milk powder or culinary powder (Bhandari et al., 2013, Kim et al., 2002, Kowalska and Lenart, 2005).

In Figure 3.6, the cumulative, volume based particle size distributions of the two carbohydrate powders in dependance of the particle diameter are presented. It can be seen that the distribution of lactose ranging approximately from 100 µm to 450 µm is broader compared to the one of sucrose ranging from approximately 200 µm to 450 µm. Thus, the lactose powder contains a finer fraction.

TABLE 3.2: Characteristic size values of sucrose and lactose.

Size values	sucrose	lactose
$d_{10,3}$ [µm]	231	156
$d_{50,3}$ [µm]	287	247
$d_{90,3}$ [µm]	355	354
d_{32} [µm]	272	214
S_v [1/mm]	26.3	37.4
Ψ [−]	0.84	0.75

Table 3.2 provides the characteristic size values for sucrose and lactose for a detailed specification. The $d_{10,3}$, $d_{50,3}$ and the Sauter diameter confirm the presence of a finer fraction in the lactose powder resulting in a higher volume specific surface area S_v. Furthermore, the particle shape of sucrose and lactose is indicated by a sphericity of 0.84 and 0.75, respectively.

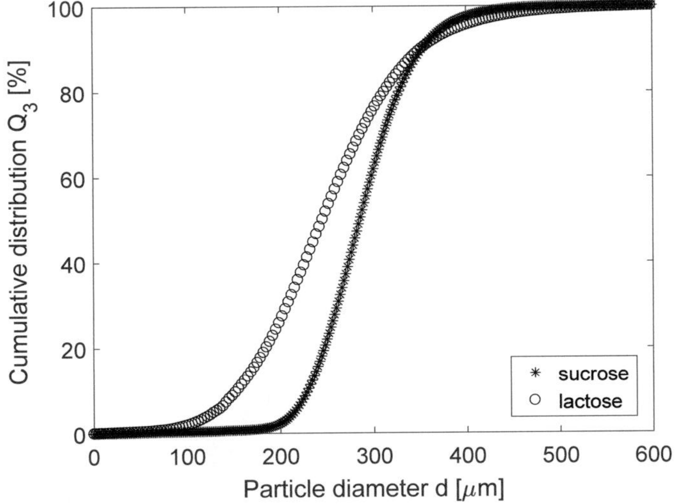

FIGURE 3.6: Cumulative particle size distribution Q_3 of sucrose and lactose.

The apparent density of sucrose and lactose was measured using a helium pycnometer (method described in section 2.1). The sucrose density with $1553 \, \mathrm{kg\,m^{-3}}$ and the lactose density with $1490 \, \mathrm{kg\,m^{-3}}$ are really close to each other.

In Figure 3.7, the capillary constant and the porosity of the two disaccharides are presented. Comparing the capillary constants in Figure 3.7 (a), it is obvious that the hexane penetration is preferred into the lactose powder. This can be explained by a larger pore space expressed by the porosity in the lactose powder. The decreased porosity of sucrose compared to lactose is a result of the higher sphericity which leads generally to higher

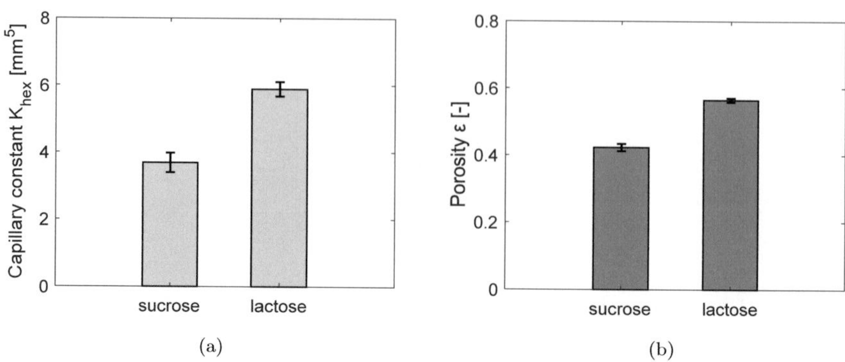

FIGURE 3.7: Capillary constant (a) and porosity (b) of sucrose and lactose powder.

packing densities (Zou and Yu, 1996). Additionally, the effective pore radii were calculated for sucrose and lactose with Eq. 4.5 which results in values of 6.8 µm and 5.7 µm, respectively.

The particle size distribution of the four size fractions of sodium chloride is shown in Figure 3.8, where s_1 is the finest, s_2 and s_3 the intermediate and s_4 the coarsest fraction. s_1 is in the range of 100 µm to 300 µm, s_2 in the range of 200 µm to 400 µm, s_3 in the range of 300 µm to 650 µm and s_4 in the range of 500 µm to 950 µm. The distribution of the fractions s_1 and s_2 are very narrow, while s_3 and s_4 show broader distribution.

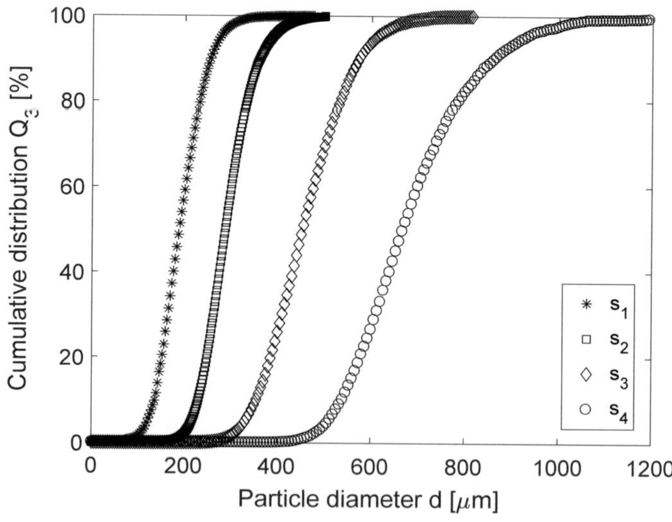

FIGURE 3.8: Cumulative particle size distribution Q_3 of four different sodium chloride size fractions.

The trend of the particle size distribution can be emphasized by considering Table 3.3. The characteristic diameters $d_{10,3}$, $d_{50,3}$, $d_{90,3}$ and d_{32} increase from s_1 to s_4, whereas the specific surface area decreases with increasing particle size. The sphericity values are not depending on the particle size, since they are all in the same size range.

TABLE 3.3: Characteristic size values of four size fractions of sodium chloride powder.

Size values	s_1	s_2	s_3	s_4
$d_{10,3}$ [µm]	154	232	358	541
$d_{50,3}$ [µm]	205	288	459	674
$d_{90,3}$ [µm]	265	362	579	869
d_{32} [µm]	171	257	444	664
S_v [1/mm]	43.4	29.2	16.9	11.3
Ψ [−]	0.81	0.80	0.80	0.80

Moreover, the four salt powders are quantified in terms of capillary constant and porosity as shown in Figure 3.9. As it was already perceivable for the sphericity of the four fractions, the porosities do not differ. Thus, an equally dense packing was achieved for the four samples. However, the capillary constants exhibit differences. The hexane penetration increases slightly from the fraction s_1 to fraction s_3 and show a sudden decrease for the coarsest fraction. Here, two effects are important to explain these phenomena. On the one hand, the penetration is increased with increasing size of the pores. On the other hand, macrovoids lead to the stop of the liquid flow and a narrow pore size distribution favours the penetration (Hapgood et al., 2002). Thus, the capillary constants from s_1 to s_3 increase, since the particle size and, therefore, the pore size increase. The lower capillary constant of s_4 is very likely due to the presence of macrovoids in the coarsest fraction.

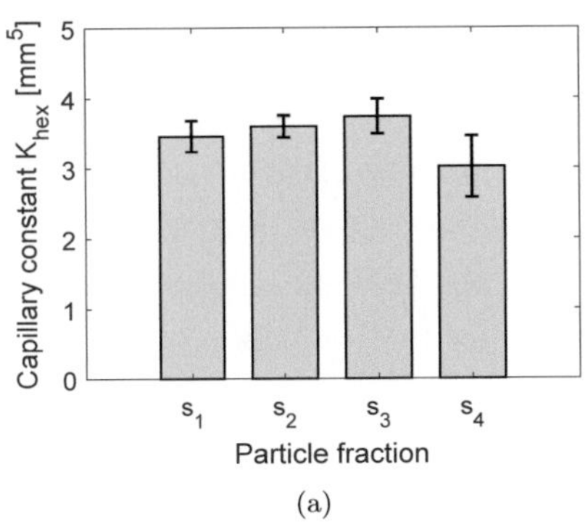
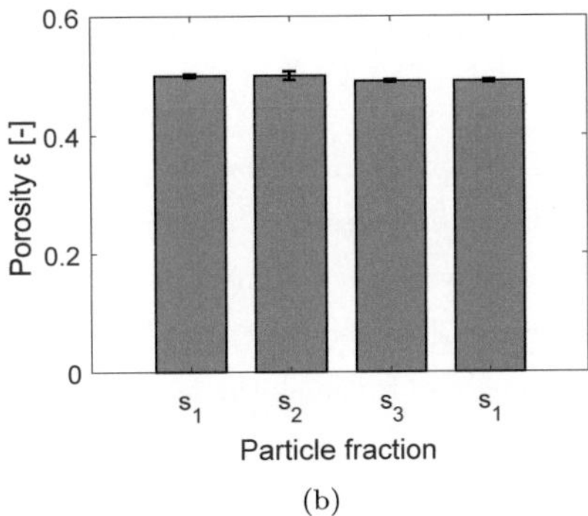

FIGURE 3.9: Capillary constant (a) and porosity (b) of four different salt fractions.

By applying Eq. (4.5), effective pore radii can be calculated from the capillary constant and the porosity. For the samples s_1, s_2, s_3 and s_4, values of 4.49 µm, 4.67 µm, 5.02 µm and $4.06\,\mu m$ are determined. Since the porosities stay almost constant, the effective pore radii follow the trend of the capillary constants. The density of sodium chloride was measured by means of a helium pycnometer, which results in a value of $2210\,kg\,m^{-3}$.

3.3.2.2 Heterogeneous food powders

The two whole milk powders were also characterized in terms of their powder properties. The particle size distribution is presented in Figure 3.10 and shows slight differences for the two powders. While the particle size of the standard whole milk powder ranges from $70\,\mu m$ to $500\,\mu m$, the WMP ff is coarser with a size from $100\,\mu m$ to $700\,\mu m$.

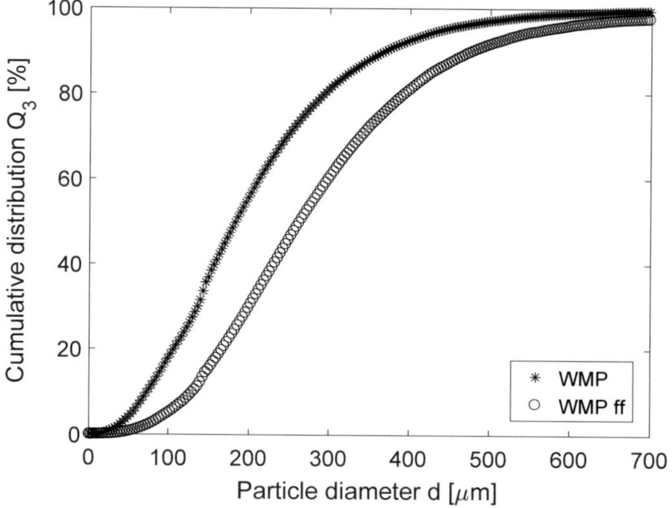

FIGURE 3.10: Cumulative particle size distribution Q_3 of standard whole milk powder (WMP) and of whole milk powder with free surface fat (WMP ff).

The characteristic size values of the milk powders are provided in Table 3.4. The specific diameters present the same trend that can be seen in the particle size distribution. Moreover, the volume specific surface area follows the trend and shows the larger value for WMP. The sphericity differs only slightly for the two milk powders and is in the same size range as the lactose powder.

The apparent density of the milk powder was measured by helium pycnometry (method described in section 2.1). Values of $1300\,kg\,m^{-3}$ and $1150\,kg\,m^{-3}$ were determined for WMP and WMP ff, respectively. It is obvious that the density for WMP ff is significantly lower than for WMP. Since WMP and WMP ff have the same composition, this difference can be only explained by the presence of more fat on the surface in case of WMP ff. It is very likely that the fat seals the surface in a better way than lactose or protein.

TABLE 3.4: Characteristic size values of WMP and WMP ff.

Size values	WMP	WMP ff
$d_{10,3}$ [µm]	75	132
$d_{50,3}$ [µm]	185	265
$d_{90,3}$ [µm]	370	480
d_{32} [µm]	135	216
S_v [1/mm]	59.6	38.4
Ψ [$-$]	0.74	0.72

Therefore, the helium gas can penetrate into pores in lactose or protein on the particle surface, but not in the fat covered particles of WMP ff which results in a lower apparent density for WMP ff.

A further characterization of powder properties of the milk powders is given in Figure 3.11. The capillary constants in Figure 3.11 (a) indicate an increasingly improved hexane penetration from WMP ff to WMP. On the other hand, the porosity in Figure 3.11 (b) shows no difference for the two powders. Therefore, even if the overall porosity is comparable for the two milk powders, the larger capillary constant for WMP imlies a more homogeneous pore network, while a decreased value for WMP ff imlies more heterogeneity due to the presence of macrovoids. The presence of macrovoids was also reported by Dang-Vu and Hupka (2009), Hapgood et al. (2002). The calculation of effective pore radii results in values of 3.0 µm and 2.3 µm for WMP and WMP ff, respectively.

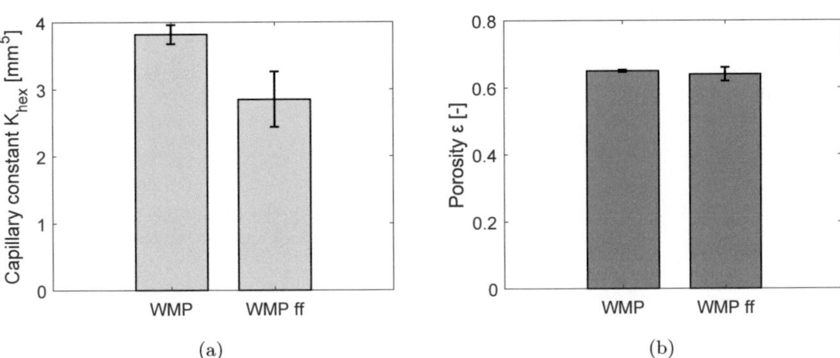

FIGURE 3.11: Capillary constant (a) and porosity (b) of WMP and WMP ff.

3.3.3 Powder properties of mixed systems

In order to improve the understanding of real food systems such as milk powder, heterogeneous model powders were produced by mixing hydrophilic and hydrophobic powders.

In a first steps two inert glass powders, a cleaned hydrophilic one and a silanized hydrophobic one, were used for the preparation of heterogeneous mixtures. Secondly, heterogeneous and soluble mixtures were created by mixing sucrose and hydrophobic glass powder and by mixing sodium chloride powder s_1 and shellac coated glass beads. Hereinafter, the two soluble and heterogeneous systems are presented more detailed, since capillary constant and porosity may change due to mixing of two differently sized and shaped powders.

3.3.3.1 Mixture of sucrose and hydrophobic glass

The mixing of sucrose powder and the silanized coarse glass powder fraction leads to a slight deviation in capillary constant and porosity. Even if the two powders are similar in their particle size distribution (compare Figure 3.4 and 3.6), a small deviation in size can impact the pore network of the mixtures. Furthermore, the different sphericities of coarse glass (0.92) and sucrose (0.84) influence the pore structure.

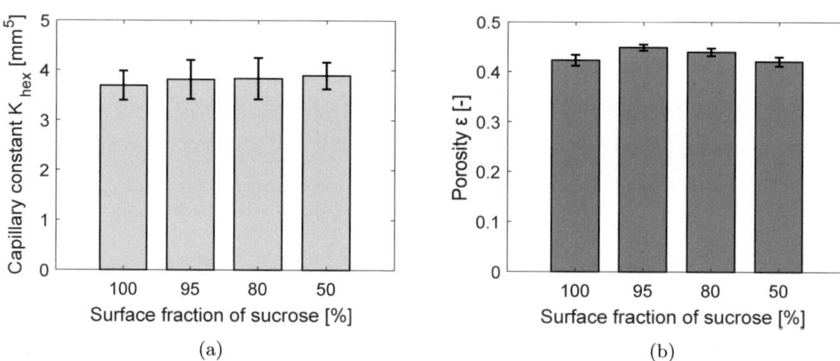

FIGURE 3.12: Capillary constant (a) and porosity (b) of sucrose-glass mixtures containing 100 %, 95 %, 80 % or 50 % of sucrose surface.

In Figure 3.12 (b), the porosity of the four mixtures shows an impact of composition. By adding 5 % of glass to sucrose, the porosity increases which means that the packing gets looser. But a further addition of glass beads with a high sphericity decreases the porosity again resulting in the lowest porosity for the 50:50-mixture of 0.42. Unlike the porosity, the capillary constants in Figure 3.12 (a) do not show a significant trend with the mixture composition due to high standard deviations. By means of the capillary constant and the porosity, effective pore radii of the packings were determined (Eq. 4.5) in order to insert the values into the model. Effective pore radii of 6.8 μm, 6.1 μm, 6.4 μm and 7.1 μm were calculated for the sucrose mixtures with increasing glass powder in the sample. It seems that the mixed systems with 5 % and 20 % of glass surface contain macro voids, while the powder beds with 0 % and 50 % of glass surface consist of a more uniform pore network. Higher amounts of smooth surface due to more glass beads and less edges due to less sucrose particles support a homogeneous and, thus, a faster uptake of hexane into the system.

3.3.3.2 Mixture of sodium chloride and shellac coated glass

In Figure 3.13 (a) and (b), the capillary constant and the porosity of mixtures prepared of sodium chloride fraction s_2 and the shellac coated coarse glass fraction are presented. Comparing the particle size distributions (Figures 3.4 and 3.8) and the characteristic size values (Tables 3.1 and 3.3) of the single components, it is obvious that the shellac coated glass fraction is larger. Considering the shape, the sphericity of sodium chloride is with 0.80 lower than the one of shellac coated glass. Mixing of both components leads to a decrease of porosity with increasing quantity of inert component in the sanmples. The lowest porosity is again achieved for the 50:50-mixture. It seems that by the addition of more spherical glass beads to the salt fraction, the packing density can be increased and, thus, the porosity.

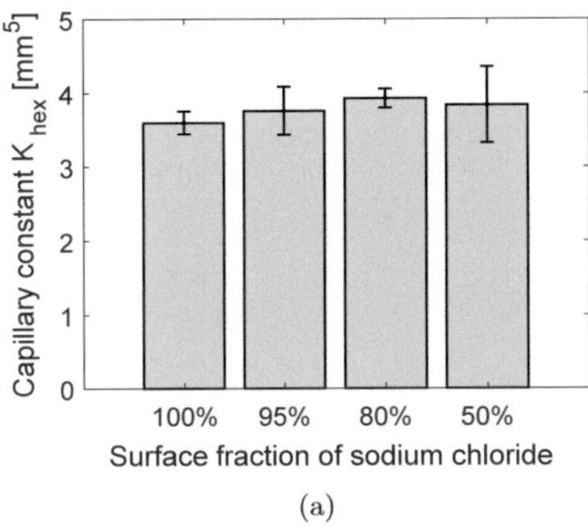
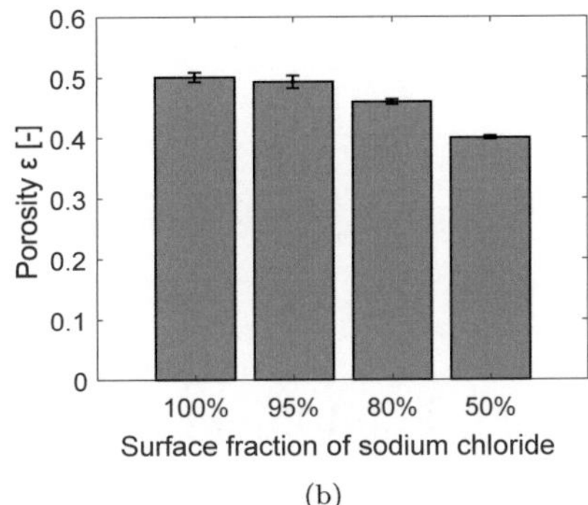

FIGURE 3.13: Capillary constant (a) and porosity (b) of mixtures with sodium chloride and shellac coated glass beads containing 100 %, 99 %, 95 %, 90 %, 80 %, 70 %, 60 % or 50 % of sodium chloride surface.

For the capillary constants, there is no dependency on the composition of sodium chloride and shellac coated glass beads observable, even if the porosity decreases from the 100 %-sample to the 50 %-sample. The condition of the pore network seems to change from large and heterogeneous to small and more homogeneous. Since large and homogeneous pores favour the penetration, the capillary constant is balanced in the four mixtures. From the capillary constant and the porosity, the effective pore radius can be calculated by using Eq. (4.5). For the samples 100 %, 95 %, 80 % and 50 %, values of 4.67 µm, 5.01 µm, 6.03 µm and 7.80 µm were determined. Since the capillary constant stays constant and the porosity decreases, the pore radius counteracts the porosity and increases with increasing quantity of inert component in the mixtures. An increasing effective pore radius does not mean that the real pore radius in the network increases.

3.4   Characterization of liquid properties

Two main liquids are important in this work. On the one hand, the standard wetting liquid which was distilled water and on the other hand n-hexane as ideal wetting liquid. Water was chosen for performing all capillary wetting experiments since water or water based liquids are generally used for the rehydration of food powders (Schubert, 1990). In order to gain information about the pore network within the investigated powders, an ideal spreading resulting in a contact angle of 0 was assumed by using n-hexane. All experiments were carried out at ambient temperature (20 °C). A summary of all important liquid properties are presented in Table 3.5. N-hexane was purchased from Carl Roth (Germany) and distilled water was obtained from the distilled water supply from university.

TABLE 3.5: Liquid properties of distilled water and n-hexane at 20 °C (Lide and Kehiaian, 1994, Vargaftik et al., 1983).

Liquids	Density ρ [kg m^{-3}]	Surface tension σ [mN m^{-1}]	Viscosity η [mPa s]
distilled water	998	72.8	1.00
n-hexane	661	18.4	0.33

3.5   Solid-liquid interactions

The characterization of inert and food materials was performed by dynamic contact angle determination. Therefore, contact angles were measured over time by applying the sessile drop method. For inert systems, contact angle determination was perfomed on equally treated glass slides instead of glass beads since the size of the beads was too small for this purpose. For the description of interactions between water and food powders, either tablets were pressed or monolayers of particles were produced.

3.5.1   Interactions in inert systems

For the hydrophilic system the contact angle was recorded for a short time period of six seconds since the dynamics is fast when the liquid droplet spreads over the hydrophilic glass slide. A series of contact angles ranging from 17.4° to 7.2° within the observed time period was measured.

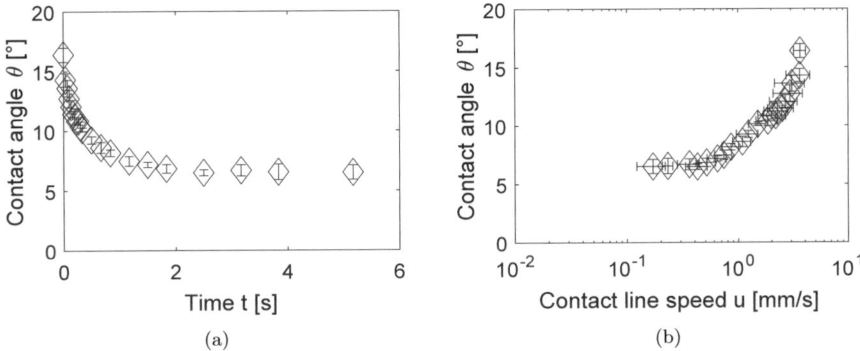

FIGURE 3.14: Dynamic contact angle on hydrophilic glass over time (a) and contact line speed (b).

In Figure 3.14 (a), it can be seen that the curve exponentially decreases over time and approaches a steady state value. In Figure 3.14 (b), the contact angle is plotted over the contact line speed. The contact line speed is defined as the distance which is covered by the motion of the three-phase point in a certain time interval during spreading of the droplet. This graph shows a decrease of contact angle with decreasing contact line speed, which is in a range of $5.2\,\mathrm{mm\,s^{-1}}$ to $0.3\,\mathrm{mm\,s^{-1}}$. This phenomenon is well known in literature and, for instance, described by Siebold et al. (2000). Hamraoui et al. (2000) introduced a friction coefficient to characterize the behavior of dynamic contact angles. This implies that higher contact angles occur at increased velocities because the liquid flow is exposed to a greater friction resistance.

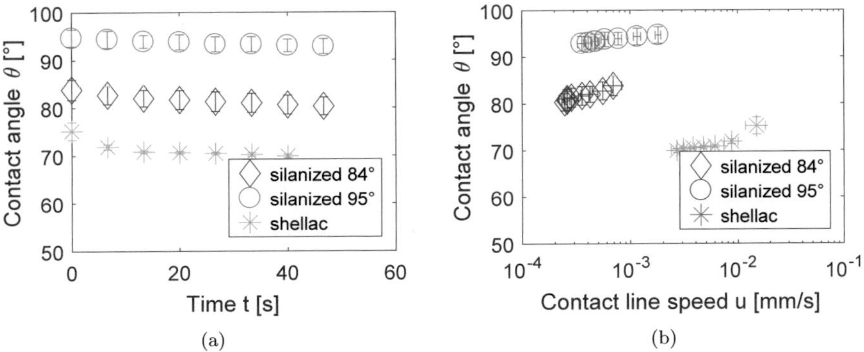

FIGURE 3.15: Dynamic contact angle on hydrophobic glass produced by silanization or shellac coating over time (a) and contact line speed (b).

The contact angles on two differently silanized ($10\,\text{mM}$ for $30\,\text{min}$ and $20\,\text{mM}$ for $72\,\text{h}$) and one shellac coated glass were captured for a longer period of 50 seconds since the motion of the three-phase point is really slow for hydrophobic materials. Even within the longer time period the alteration of the contact angles is only from $83.8°$ to $80.3°$ and from $94.7°$ to $92.9°$ for the two silanized glass slides and from $75.2°$ to $69.9°$ for the shellac coated glass slide, which can be seen in Figure 3.15 (a). Hereinafter, the glass which was silanized with a concentration of $10\,\text{mM}$ for $30\,\text{min}$ is labeled as silanized $84°$ and the glass with silanization parameter of $20\,\text{mM}$ for $72\,\text{h}$ is labeled as silanized $95°$. The small alteration in contact angle results in very slow contact line speeds of $0.7 \times 10^{-3}\,\text{mm}\,\text{s}^{-1}$ to $0.2 \times 10^{-3}\,\text{mm}\,\text{s}^{-1}$ (silanized $84°$), of $1.8 \times 10^{-3}\,\text{mm}\,\text{s}^{-1}$ to $3.6 \times 10^{-4}\,\text{mm}\,\text{s}^{-1}$ (silanized $95°$) and $1.5 \times 10^{-2}\,\text{mm}\,\text{s}^{-1}$ to $2.7 \times 10^{-3}\,\text{mm}\,\text{s}^{-1}$ (shellac coated glass) as presented in Figure 3.15 (b). Thus, it can be concluded that both, the silanization procedure and the coating step, lead to a significant alteration of the glass surface towards increased hydrophobicity. Since both treatments results in a contact angle above the critical one of $55°$ decribed by Raux et al. (2013), the two silanized and shellac coated glass beads can be used as hydrophobic powder for the preparation of heterogeneous mixtures.

The impact of surface roughness is neglected since the SEM-image of glass material showed a smooth appearance at a magnification of 500 times (Figure 3.1 (b)). Palzer et al. (2001) measured a surface roughness of about $0.03\,\mu\text{m}$ on microscopic glass slides as they were used in this thesis. Furthermore, Busscher et al. (1984) figured out that there is no influence of surface roughness on the contact angle as long as the roughness is below $0.1\,\mu\text{m}$.

Since hydrophilic and hydrophobic glass powders were mixed in different compositions ranging from $100\,\%$ of hydrophilic glass to $50\,\%$ hydrophilic plus $50\,\%$ hydrophobic glass powder, Cassie-Baxter contact angles describing the heterogeneous wetting behaviour were determined. The single components are characterized by dynamic contact angle of $17.4°$ and $83.8°$ for hydrophilic and hydrophobic glass, respectively. The mixed contact angles of the mixtures and the cosine of the angles are presented in Table 3.6. The values ranges from $19.0°$ for the 99:1-mixture to $57.9°$ for the 50:50-mixture.

3.5.2 Interactions in food systems

In this subsection, the analysis of interactions between the food materials and water is described in detail.

TABLE 3.6: Cassie-Baxter contact angles for mixtures of hydrophilic and hydrophobic glass powder.

Surface fraction hydrophilic glass f_1 [%]	Surface fraction hydrophobic glass f_2 [%]	Cassie-Baxter contact angle θ_{CB} [°]	$\cos\theta_{CB}$ [-]
100	0	17.4	0.95
99	1	19.0	0.95
95	5	24.2	0.91
90	10	29.6	0.87
80	20	38.3	0.78
70	30	45.5	0.70
60	40	52.0	0.62
50	50	57.9	0.53

3.5.2.1 Homogeneous food powders

For describing the interactions between water and the homogeneous food powders sucrose, lactose and sodium chloride, tablets were pressed from primary particles as explained in subsection 2.2.5 and used as solid surface for the sessile drop method. As the pressing of powders does not create smooth surfaces, roughness of the tablets was investigated as described in subsection 2.2.7.

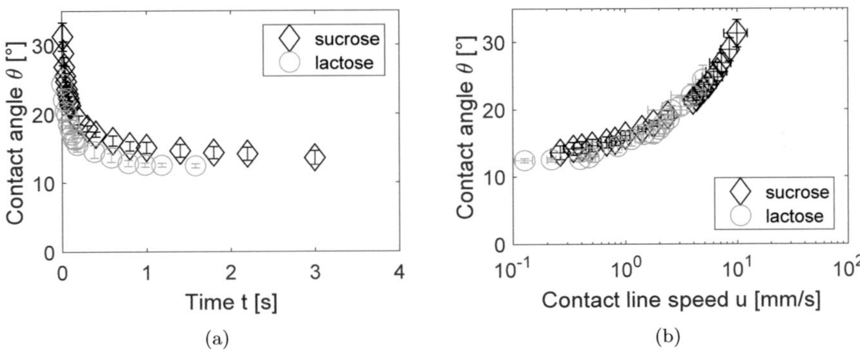

FIGURE 3.16: Dynamic contact angle on sucrose and on lactose tablet over time (a) and contact line speed (b).

Figure 3.16 presents the water contact angle in dependance of time (a) and of contact line speed (b) for sucrose and lactose. For both disaccarides, a hydrophilic behaviour can be observed. The contact angle of sucrose and lactose spreads from 31.3° to 13.6° within three seconds and from 24.4° to 12.5° within two seconds, respectively (Figure 3.16 (a)). Since the volume stayed only constant during the first three (sucrose) or two (lactose) seconds the contact angle was analyzed within these time periods. The curve approximates a constant contact angle for both materials. However, it is not possible to measure a static contact angle for these systems due to dissolution and movement of the contact line. For capillary rise experiments as investigated in this study the dynamic contact angle is more relevant than the static one. The dynamics of the contact angle can be better shown in Figure 3.16 (b), where the contact angle is plotted over the contact line speed. The contact angle decreases with decreasing contact line speed. The velocities of the contact line range from $10\,\mathrm{mm\,s^{-1}}$ to $0.3\,\mathrm{mm\,s^{-1}}$ within the observed period for sucrose and from $4.9\,\mathrm{mm\,s^{-1}}$ to $0.1\,\mathrm{mm\,s^{-1}}$ for lactose. Dupas et al. (2017) measured a water contact angle of 15.8° on crystalline sucrose at a contact line speed of $0.01\,\mathrm{mm\,s^{-1}}$, which almost corresponds to static conditions. But they do not mention if roughness was considered during the contact angle determination. For lactose, a value of 30° is given in literature for the static case (Lerk et al., 1976). Dynamic contact angles of amorphous lactose were presented by Dupas (2012) ranging from 11° to 2° for contact line speeds from $0.34\,\mathrm{mm\,s^{-1}}$ to $4 \times 10^{-4}\,\mathrm{mm\,s^{-1}}$. The deviation of the contact angle values can be explained by the state of lactose. While in this thesis lactose in a mainly crystalline state was used, Dupas (2012) performed the experiments with amorphous lactose.

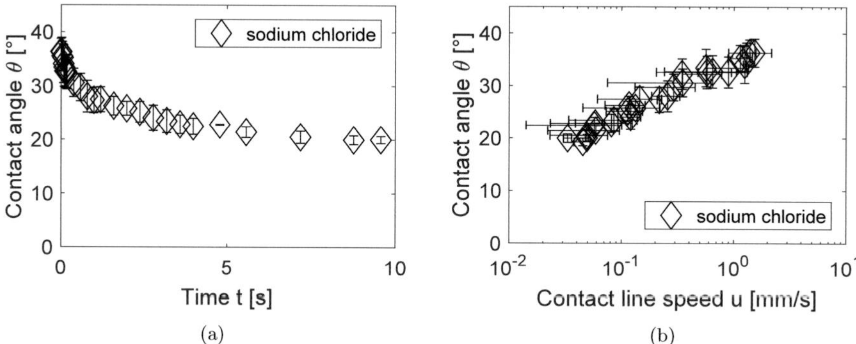

FIGURE 3.17: Dynamic contact angle on sodium chloride tablet over time (a) and contact line speed (b).

Figure 3.17 shows the dynamic contact angle of sodium chloride which is the third homogeneous food powder used in this thesis. Sodium chloride is also characterized by a hydrophilic behaviour. The contact angle spreads from 36.3° to 19.2° within ten seconds (Figure 3.17 (a)) which results in contact line speed ranging from $1.4\,\mathrm{mm\,s^{-1}}$ to $0.05\,\mathrm{mm\,s^{-1}}$. In literature, a static contact angle of 28° for sodium chloride is given (Lerk et al., 1976).

Roughness has an impact on the contact angles which are determined by sessile drop method. Thus, the tablet roughness was measured for sucrose, lactose and sodium chloride and is shown in Table 3.7.

TABLE 3.7: Surface roughness of food powder tablets.

	Sucrose	Lactose	Sodium chloride
Roughness R_a [µm]	1.99 ± 0.16	1.60 ± 0.12	1.09 ± 0.16

As it is described in subsection 1.1.1, static contact angles below 90° decrease and static contact angles above 90° increase with increasing roughness (Wenzel, 1936). In this thesis, the contact angles were determined dynamically and, hence, the influence of roughness on static contact angle may not help for interpretation. Considering the observations of Palzer et al. (2001) for the dynamic case, it can be expected that the dynamic contact angle on sucrose, lactose and sodium chloride is overestimated due to the roughness of the tablet. However, during capillary rise experiments the water has to pass through a network of food particles, which also act as rough surface influencing the solid-liquid interactions. Therefore, the speed dependent contact angles on rough surfaces were used for penetration investigations.

3.5.2.2 Heterogeneous food powders

Multicomponent food powders were characterized by manufacturing of thin layers of the primary particles as described in subsection 2.2.6 and conducting the sessile drop measurement. Due to the presence and distribution of different components in multicomponent food powders, a pressing step would alter the surface composition of the particles and thus it is not suitable.

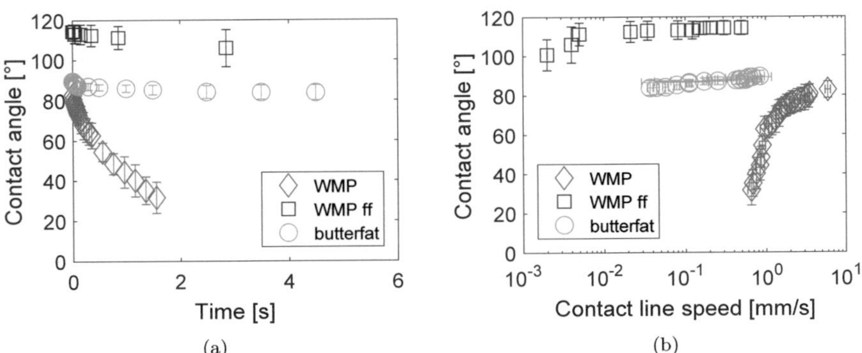

FIGURE 3.18: Dynamic contact angle as a function of time (a) and of contact line speed (b) for WMP, WMP ff and butterfat.

In Figure 3.18, the dynamic contact angles on two milk powders and butterfat are presented. It can be immediately seen that the contact angles in dependence of time (a) and in dependence of contact line speed (b) differ only for WMP, while WMP ff and butterfat stay nearly constant. Comparing WMP and WMP ff which have the same composition of protein, lactose and fat the completely different interactions with water are noteworthy. While the contact angle on WMP decreases from 82.7° to 31.6° within 1.6 s, the contact angle on WMP ff is clearly larger and changes only slighty from 114.4° to 100.7° within 7.9 s. The contact line speed measured for WMP ranges from $6.0\,\mathrm{mm\,s^{-1}}$ to $0.7\,\mathrm{mm\,s^{-1}}$ and for WMP ff from $0.5\,\mathrm{mm\,s^{-1}}$ to $2.0 \times 10^{-3}\,\mathrm{mm\,s^{-1}}$. The differences can be explained by the loction of the hydrophobic components in the particles. In the WMP ff powder the milk fat is accumulated on the surface due to a missing homogenization step. Thus, the contact angle of water on milk fat is determined when analyzing the WMP ff powder. On contrary, a WMP particle is heterogeneous in surface composition with lactose, protein and fat. Therefore, the contact angle decrease occurs due to spreading on hydrophilic components and dissolution of lactose. Butterfat is the main fatty component in milk fat and, thus, respresents the most hydrophobic ingredient in the milk powders. When comparing the contact angle of WMP ff and butterfat it is obvious that the higher contact angle of WMP ff can be only explained by roughness effects. Since the contact angle on butterfat is the maximum contact angle a milk powder can achieve on smooth surfaces, the roughness of the monolayer of particles within the thin layer leads to the enlarged contact angle. Silva and O'Mahony (2017) measured dynamic contact angles on a whole milk powder tablet decreasing from 95° to 80° within 5 s, respectively. The measured contact angle in the study of Silva and O'Mahony (2017) are comparable to the contact angle on butterfat in this work. Thus, it seems that pressing of the tablet resulted in the accumulation of the fatty component on the surface.

3.5.3 Interactions in mixcd systems

For description of water interactions with heterogeneous powder mixtures containing two materials with different contact angles, the Cassie-Baxter equation is applied (Eq. 1.6). The Cassie-Baxter contact angle helps to understand the overall behaviour of the heterogeneous system. The values are later used for modelling the liquid penetration into heterogeneous sytem.

3.5.3.1 Sucrose and hydrophobic glass mixtures

One heterogeneous system in this thesis is created by mixing sucrose and silanized glass beads with dynamic contact angles of 31.3° and 83.8°, respectively. The resulting Cassie Baxter contact angles and the cosines of the contact angles are provided in Table 3.8. It can be seen that the contact angles of the mixtures ranges from 35.2° for the 95:5-mixture to 61.2° for the most hydrophobic mixture.

TABLE 3.8: Cassie-Baxter contact angles for mixtures of sucrose and silanized glass.

Surface fraction sucrose f_1 [%]	Surface fraction silan. glass f_2 [%]	Cassie-Baxter contact angle θ_{CB} [°]	$\cos\theta_{CB}$ [-]
100	0	31.3	0.85
95	5	35.2	0.82
80	20	45.2	0.71
50	50	61.2	0.48

3.5.3.2 Sodium chloride and shellac coated glass mixtures

Secondly, hydrophilic sodium chloride powder and hydrophobic, shellac coated glass beads were mixed in different concentrations ranging from 100 % of sodium chloride to 50 % sodium chloride plus 50 % shellac coated glass powder. The single components are characterized by dynamic contact angles of 36.3° and 75.2° for sodium chloride and shellac, respectively. The Cassie-Baxter contact angles of the mixtures are presented in Table 3.9. The values ranges from 38.9° for the 99:1-mixture to 57.9° for the 50:50-mixture.

TABLE 3.9: Cassie-Baxter contact angles for mixtures of sodium chloride and shellac coated glass.

Surface fraction sodium chloride f_1 [%]	Surface fraction shellac glass f_2 [%]	Cassie-Baxter contact angle θ_{CB} [°]	$\cos\theta_{CB}$ [-]
100	0	36.3	0.81
95	5	38.9	0.78
80	20	45.9	0.70
50	50	57.9	0.53

4

Model description

For the prediction of capillary wetting into a heterogeneous system, two different models were applied in this work. The liquid penetration into a heterogeneous, inert system was modelled by an available approach from literature, whereas, a model dealing with capillary penetration into a heterogeneous system and dissolution of one component at the same time was developed.

4.1 Model for inert, heterogeneous system

For the modelling of the inert case, an approach of Benavente et al. (2002) was chosen and slightly modified to calculate theoretical penetration rates. This approach is based on the well known Washburn equation (Eq. 1.15) (Washburn, 1921). In order to convert the Washburn equation into its mass related version for a pore network, Eq. (1.15) is multiplied by the cross-sectional area of the Washburn cylinder A, the porosity of the packed bed ε and the density of the liquid ρ_l. Therefore, $M_l(t)$ is the time depending mass of liquid in the pore network of the packed bed (Eq. 4.1):

$$M_l(t) = A \cdot \varepsilon \cdot \rho_l \cdot h(t). \tag{4.1}$$

By division through the cross-sectional area A we receive the area independent penetration rate C_{theo} of liquid into the packed bed (Eq. 4.2 and 4.3):

$$C_{theo} = \frac{M_l(t)}{A \cdot \sqrt{t}}, \tag{4.2}$$

$$C_{theo} = \varepsilon \cdot \rho_l \cdot \sqrt{\frac{r_p \cdot \sigma \cdot \cos\theta}{2 \cdot \eta}}. \tag{4.3}$$

Since the inert powders contain hydrophilic as well as hydrophobic glass beads, the system cannot be described by only one contact angle. Hence, the Cassie-Baxter equation is applied to calculate apparent mixed contact angles θ_{CB} for heterogeneous systems (Eq. 1.6) where θ_1 and θ_2 represent the contact angles of the hydrophilic and the hydrophobic glass material, respectively. f_1 and f_2 are the surface proportions of the two different glass powders within the system. By inserting the Cassie-Baxter contact angle into Eq. (4.3), the slightly modified model equation (Eq. 4.4) is obtained:

$$C_{theo} = \varepsilon \cdot \rho_l \cdot \sqrt{\frac{r_p \cdot \sigma \cdot \cos\theta_{CB}}{2 \cdot \eta}}. \tag{4.4}$$

The pore radius r_p of the system can be approximated by an effective pore radius $r_{p,eff}$ The $r_{p,eff}$-value of the pore network can be calculated by using the capillary constant K_{hex}, which is experimentally determined with the Washburn setup and hexane as wetting liquid, the cross-sectional area and the porosity (Eq. 4.5). $r_{p,eff}$ is inserted in Eq. (4.4) for the pore radius r_p:

$$r_{p,eff} = \frac{2 \cdot K_{hex}}{A^2 \cdot \varepsilon^2}. \tag{4.5}$$

4.2 Model for soluble, heterogeneous food system

For predicting penetration rates into a food system containing soluble particles, it is required to consider the effect of solubility by developing a model. Dissolution of food particles during the penetration process leads to changing liquid properties, such as density, surface tension and viscosity (Hogekamp and Schubert, 2003). Furthermore, due to the shrinkage of the particles, the pore volume increases and the surface area decreases. The two latter influencing factors caused by particle shrinkage are neglected in our model since the focus was put on the first few seconds of the penetration step. It is assumed that both effects, the decrease of surface area and the increase of pore volume, are marginal compared to the changing liquid properties within the first seconds, which is discussed in detail in chapter 6. Secondly, the effect of heterogeneity in terms of contact angles has to be considered. This is realized by using Eq. (1.6) to calculate an apparent mixed contact angle of the system depending on the content of hydrophilic and hydrophobic surface. Figure 4.1 provides an overview about the modelling approach. A detailed explanation about the operating principle follows.

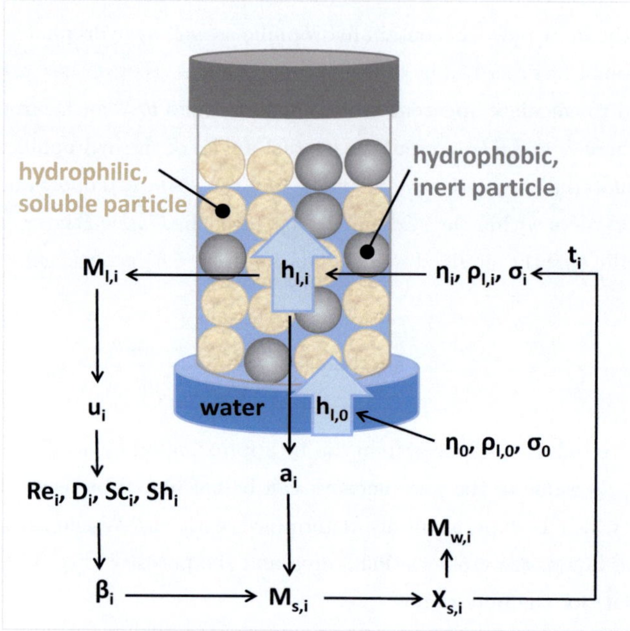

FIGURE 4.1: Schematic model of liquid penetration into mixed system.

The model was developed in a former study (Kammerhofer et al., 2018b). It is based on two differential equations, the differential mass transfer equation (Eq. 4.6) and the differential Washburn equation (Eq. 4.7). A coupled system of both differential equations was implemented, which is solved by means of the solver ode45 in MATLAB. The dissolution is described by Eq. (4.6), where M_s is the mass of dissolved component, β is the mass transfer coefficient and ρ_{sat} the density of a saturated solution. As already mentioned above, the Cassie-Baxter equation (Eq. 1.6) is applied to determine apparent, mixed contact angles. Therefore, θ_{CB} is inserted for θ in Eq. (4.7):

$$\frac{dM_s}{dt} = \beta \cdot a \cdot (\rho_{sat} - \rho_l), \tag{4.6}$$

$$\frac{dh}{dt} = \frac{r_p \cdot \sigma \cos\theta}{4 \cdot \eta \cdot h}. \tag{4.7}$$

At $t = 0$ the model starts to run with the liquid properties of water (section 3.4). Eq. (4.7) determines for each time step a penetration height h. This parameter is used to calculate the penetration mass of liquid M_l (water and dissolved component) (Eq. 4.8) and the velocity of the rise u (Eq. 4.9):

$$\frac{dM_l}{dt} = \varepsilon \cdot \rho_l \cdot A \cdot \frac{dh}{dt}, \tag{4.8}$$

$$u = \frac{dh}{dt}. \tag{4.9}$$

The velocity u is necessary to determine the Reynolds number Re (Eq. 4.10) where d_{50} represents the average diameter of the soluble particles in the powder bed:

$$Re = \frac{u \cdot d_{50} \cdot \rho_l}{\eta}. \tag{4.10}$$

The determination of Reynolds number, Sherwood number Sh and Schmidt number Sc is necessary in order to define the mass transfer coefficient β for each step (Eq. 4.11). Furthermore, β depends on the diffusion coefficient D, which is calculated with the Stokes-Einstein equation (Eq. 4.12) where k_B is the Boltzmann constant with a value of $1.38 \times 10^{-23}\,\mathrm{J\,K^{-1}}$ (Lide and Kehiaian, 1994), T is the ambient temperature $20\,^\circ\mathrm{C}$ and R_0 is the hydrodynamic radius:

$$\beta = \frac{Sh \cdot D}{d_{50}}, \tag{4.11}$$

$$D = \frac{k_B \cdot T}{6 \cdot \pi \cdot \eta \cdot R_0}. \tag{4.12}$$

The Sherwood number is estimated by the correlation for flow through packed beds of solids (Eq. 4.13) where Sh_{sphere} is the Sherwood number for a flow over single spheres and f_a is a form factor (VDI, 2013). Eq. (4.14) shows the correlation for the form factor if the bed consists of spheres of the same size (VDI, 2013):

$$Sh_{bulk} = f_a \cdot Sh_{sphere}, \tag{4.13}$$

$$f_a = 1 + 1.5 \cdot (1 - \varepsilon). \tag{4.14}$$

The Sherwood number Sh_{sphere} results from Eq. (4.14) where Sh_{lam} and Sh_{turb} are the Sherwood numbers for a laminar and a turbulent flow, respectively. With Eq. (4.16) and Eq. (4.17), Sh_{lam} and Sh_{turb} are calculated by inserting Re (Eq. 4.10) and Sc (Eq. 4.18):

$$Sh_{sphere} = 2 + (Sh_{lam}^2 + Sh_{turb}^2)^{1/2},\tag{4.15}$$

$$Sh_{lam} = 0.664 \cdot Re^{1/2} \cdot Sc^{1/3},\tag{4.16}$$

$$Sh_{turb} = \frac{0.037 \cdot Re^{0.8} \cdot Sc}{1 + 2.443 \cdot Re^{-0.1} \cdot (Sc^{2/3} - 1)},\tag{4.17}$$

$$Sc = \frac{\eta}{\rho_l \cdot D}.\tag{4.18}$$

The penetration height in the packed bed is also used to determine the mass transfer area a (Eq. 4.19), which depends on the height-independent mass transfer area a_h (Eq. 4.20). This parameter is derived by the total surface area of the particles divided by the height of the packed bed. f_1 gives the proportion of soluble component in the system:

$$a = a_h \cdot f_1 \cdot h,\tag{4.19}$$

$$a_h = \pi \cdot (1 - \varepsilon) \cdot \frac{3}{2} \cdot \frac{d_{cap}^2}{d_{50}}.\tag{4.20}$$

Eq. (4.6) provides us the mass of sucrose M_s, Eq. (4.8) the mass of liquid M_l and Eq. (4.21) calculates the load of the soluble component in the liquid for each time step. With Eq. (4.22) the mass of water can be determined:

$$X_s = \frac{\frac{M_s}{M_l}}{1 - \frac{M_s}{M_l}},\tag{4.21}$$

$$M_w = \frac{M_s}{X_s}.\tag{4.22}$$

The sucrose load of the liquid is used to determine the liquid properties. The correlation between the load of each soluble component and the solution properties are presented with Eq. (4.23), Eq. (4.24) and Eq. (4.25):

$$\rho_l = f(X_s), \tag{4.23}$$

$$\eta = f(X_s), \tag{4.24}$$

$$\sigma = f(X_s). \tag{4.25}$$

$$\rho_l = f(X_s), \tag{4.23}$$

5

Capillary wetting into inert systems

5.1 Liquid penetration into single gaps

In homogeneous and heterogeneous single gaps, experiments were performed in order to investigate the impact of contact angle on the capillary rise in an ideally straight pore. Thus, two hydrophilic glass slides or one hydrophilic and one hydrophobic glass slide were used to build up a homogeneous system or heterogeneous systems, respectively. The hydrophilic glass slide is characterized by a dynamic contact angle of $17.4°$ as determined in section 3.5, while the heterogeneous systems consist of the hydrophilic glass slide and one of the two silanized, hydrophobic glass slides ($83.3°$ or $94.7°$). For the heterogeneous systems, the resulting Cassie Baxter contact angles are $57.9°$ and $63.8°$, respectively. The values for the Cassie-Baxter contact angles were taken from Table 3.6 (50-50-mixtures).

The equilibrium height h_{eq} of water in the three different setups was determined in dependency on the gap width w. The gap widths were set to be $0.47\,\text{mm}$, $0.82\,\text{mm}$ and $1.17\,\text{mm}$. The measured data points were compared with theoretical equilibrium heights which were calculated by using Eq.1.19. The surface tension and the density of water were taken from Table 3.5 and the gravitational acceleration is $9.81\,\text{m s}^{-2}$. In Figure 5.1, the measured equilibrium heights of the three different systems are plotted versus the gap width. As expected, it can be seen that the equilibrium height decreases with increasing gap width. The lines present the theoretical heights depending on the gap width. Generally, a good agreement between experiment and theory was obtained for the three systems. The model overestimates the equilibrium heights for the hydrophilic system and for one heterogeneous system ($17°+84°$). This deviation can be due to impurities on the hydrophilic glass slides which lowers immediatley the contact angle. The model tends to slightly overestimate the equilibrium height at large gaps and underestimate it for small gaps in case of the other heterogeneous system ($17°+95°$).

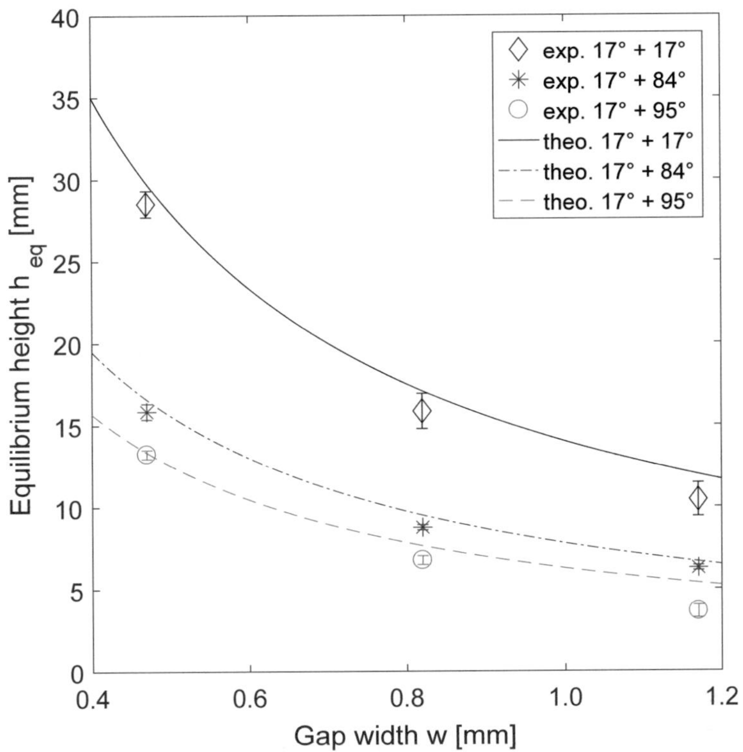

FIGURE 5.1: Experimental and theoretical equilibrium heights in single gap depending on gap width for two hydrophilic walls ($17°+17°$) and for two systems with one hydrophilic and one hydrophobic wall ($17°+84°$ and $17°+95°$).

As already described in subsection 1.1.2, O'Brien et al. (1968) studied the capillary rise of different liquids between dissimilar solid plates and evaluated their model equation with experimental data. They measured the equilibrium heights in heterogeneous systems, such as glass-Teflon or glass-acrylic with contact angles of the materials of 14°, 74° and 110° for glass, acrylic and Teflon, respectively. In order to compare their measured data with the theoretical rise, they used the factor wh which is the product of the gap width multiplied by the equilibrium height. The deviation of the mean observed values from the predicted values is 2 % for the acrylic-glass system and 21 % for the Teflon-glass system. However, by using the same comparison factor for the homogeneous and heterogeneous glass systems maximum deviations of 8 %, 6 % and 15 % for the hydrophilic system, heterogeneous system $17°+84°$ and heterogeneous system $17°+95°$, respectively, are received. Nevertheless, it can be concluded that it is possible to predict equilibrium heights in heterogeneous single gaps using the mixed dynamic contact angles calculated with the Cassie-Baxter equation.

The results which were obtained in the single gap setup provide information about the water penetration into single gaps with heterogeneous walls (hydrophilic and hydrophobic) but with a constant geometry. The knowledge about this simplified wetting process on a micro scale helps to understand the highly complex wetting process of a powder containing, besides heterogeneous surfaces, additionally a heterogeneous pore network. From this section of the study, it can be concluded that the dynamic contact angles of both walls, the hydrophilic and the hydrophobic, act equally on the capillary rise with a constant pore size.

5.2 Liquid penetration into powder systems

After the analysis of wetting experiments in homogeneous and heterogeneous single gaps, the focus was set on the water penetration into complex pore networks within heterogeneous powder. Therefore, dynamic liquid penetration into heterogeneous mixtures of coarse and fine glass powders were measured using the Washburn tool of the K100 tensiometer as described in section 2.5. A detailed characterization of the powder properties of the mixtures is provided in subsection 3.3.1 and a summary about the solid liquid interactions can be found in subsection 3.5.1.

5.2.1 Influence of heterogeneous, coarse powder on penetration rate

The determination of experimental penetration rates into heterogeneous systems is based on the squared mass-time plots, which are presented in Figure 5.2. The graphs show the water uptake over time into the particle packing of different compositions of the hydrophilic powder and the hydrophobic (84°) powder. The fastest penetration can be observed for the pure hydrophilic sample, whereas the slowest rate is obtained for the 50:50-mixture. While the squared mass-time-plot of the 100 % hydrophilic sample shows the typical Washburn behavior with a linear increase and a subsequent significant change of the slope, with higher contents of hydrophobic material, the intensity of the change of the slope gets less. Therefore, the linear increase gets shorter for the more hydrophobic samples. In order to compare the penetration rates, an experimental penetration rate C_{exp} is defined which is the slope of the linear part of the squared mass-time-plots divided by the cross-sectional area of the Washburn cylinder A. A coefficient of determination R^2 of at least 0.98 is set for selection of the linear part and its slope. The chosen time interval ranges from 0 s to 6 s (or longer). The different final squared mass-values of the samples 100 % to 70 % are due to slightly different total packing heights, which do not influence the penetration rate. A higher hydrophobic content in the 60 %- and 50 %-samples results in a lower equilibrium level of the liquid rise. In these samples the full saturation of the packing is not achieved.

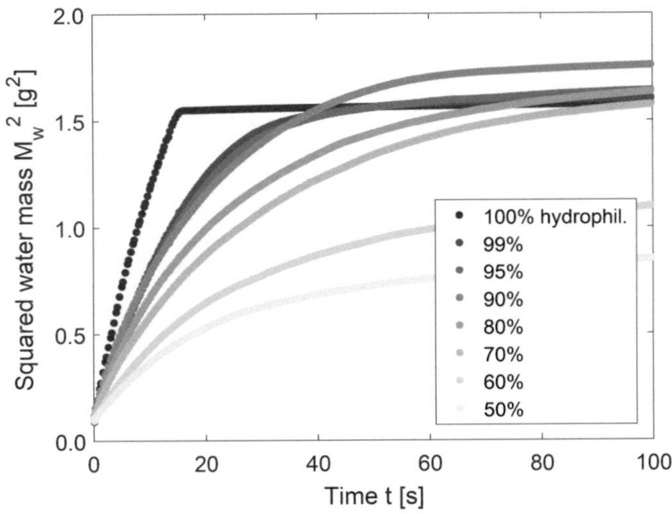

FIGURE 5.2: Squared mass of penetrated water over time into mixtures of hydrophilic and hydrophobic (84°) coarse glass powder.

The experimental penetration rates for the eight different compositions of hydrophilic and hydrophobic (84°) glass powders and additional of three mixtures of hydrophilic and more hydrophobic (95°) glass powders are presented in Figure 5.3 (a). As expected, the penetration rates decrease with an increasing amount of hydrophobic powder in the samples. For a better illustration of the hydrophobicity of the samples, the penetration rates were also plotted over the the cosine of the Cassie-Baxter contact angle (Figure 5.3 (b)). The Cassie-Baxter contact angles and the cosine values are summarized in Table 3.6. The trend of (a) is confirmed in (b) since decreasing penetration rates correlate with decreasing cosine values which corresponds to increasing contact angles. For the most hydrophilic system ($\cos \theta_{CB} = 0.95$), a jump can be observed in the overall increasing trend of C_{exp} versus $\cos \theta_{CB}$. It seems that the addition of very small amounts of the hydrophobic component (1 % of hydrophobic surface) already leads to a significant change in the wetting behavior and, thus, in a sharp decrease in penetration rate. A further increase of the hydrophobic component in the sample results in a continuous decrease of the penetration rates. This behavior was already described by Kammerhofer et al. (2018a). It is very likely that by adding 1 % of hydrophobic surface to the system a critical concentration is achieved, which entails a different wetting behavior. A certain amount of pores appear to be impacted by the hydrophobic beads dominating the penetration rates. A critical concentration of hydrophobic material which changes the wetting behavior was also reported by Nguyen et al. (2009) and Charles-Williams et al. (2013).

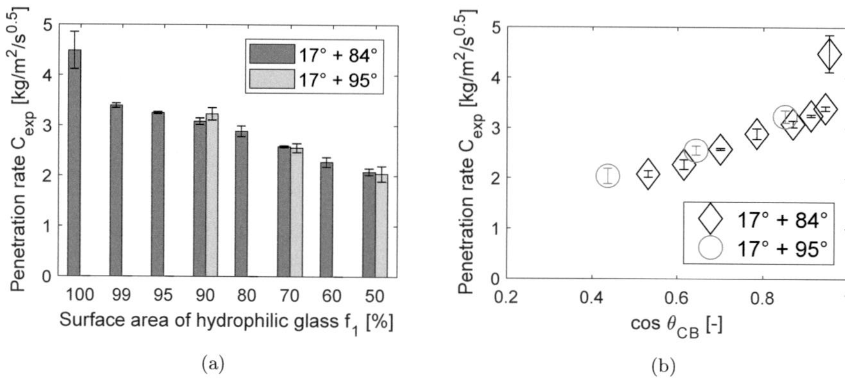

FIGURE 5.3: Experimental penetration rates C_{exp} over the surface area of hydrophilic glass (a) and over $\cos\theta_{CB}$ (b) for the coarse mixtures of $17° + 84°$ and of $17° + 95°$.

Comparing the two different hydrophobic powders with a dynamic contact angle of 84° and of 95°, it is obvious that the higher contact angle in the mixtures 90:10, 70:30 and 50:50 do not lead to a decreased penetration rate (Figure 5.3 (a)). Considering the equilibrium heights in heterogeneous single gaps, an influence of the hydrophobic contact angle could be observed. Thus, the structure of the pore network must influence the liquid penetration as it was also reported by Raux et al. (2013). Due to changes in pore shape and orientation, the critical contact angle leading to pore blockage is reduced compared to straight pores. Therefore, pores which are affected by a critical fraction of hydrophobic surfaces in the powder get blocked and are not longer available for the penetration flow. Consequently, a further increased hydrophobic contact angle does not have an influence on the wetting performance since blocked pores stay blocked. The main conclusion can be drawn that pore heterogeneity besides material heterogeneity in a powder system increases the complexity of the wetting process.

5.2.2 Influence of heterogeneous, fine powder on penetration rate

The same penetration experiments were also performed using the fine glass powder in order to investigate the effect of the particle and, hence, the pore size on the wetting behaviour. Figure 5.4 shows the squared mass over time plots for the eight different mixtures of the fine powder where a similar trend as for the coarse powder can be seen.

With increasing addition of hydrophobic glass powder the initial slope of the graphs decreases. The slight deviations of the final mass of water in the samples can be again explained by the small differences of the total packing height which does not affect the slope of the graphs. It can be observed that also the mixtures with higher concentrations of hydrophobic surface (60 %-sample and 50 %-sample) tend to achieve a similar final height as the more hydrophilic samples. This difference to the coarse powder is explicable by the pore network which changes with altering particle sizes. Due to smaller pores in the fine powder, the resulting equilibrium height of the liquid is higher which is also confirmed by Eq. (1.18).

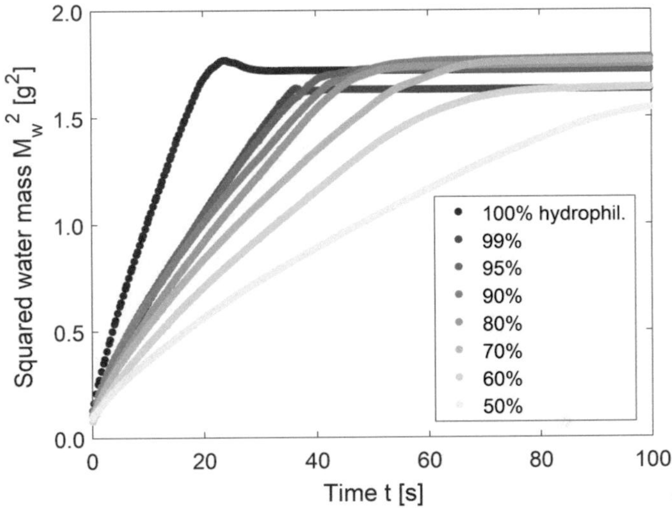

FIGURE 5.4: Squared mass of penetrated water over time into mixtures of hydrophilic and hydrophobic fine glass powder.

The experimental penetration rates which were determined identically as for the coarse powder are presented in Figure 5.5. With an increasing amount of hydrophobic surface, the same trends as for the coarse powder can be observed. By increasing the hydrophobic contact angle from 84° to 95°, there is no influence on the penetration rates for the 90:10- and 70:30-sample (Figure 5.5 (a)). But the 50:50-mixture is affected resulting in a reduced penetration rate. It seems that in the fine system, the hydrophobic contact angle gets dominant at higher concentrations. Generally, three sections can be distinguished: firstly, the low critical concentration of hydrophobic material (1 %) which changes the wetting performance significantly, secondly, a broad concentration range where a further addition of hydrophobic surface reduces the penentration rate only slightly and an increased hydrophobic contact angle has no effect (until about 80 % of hydrophilic powder) and thirdly, at high concentration of hydrophobic material where the penetration rate decreases significantly with the concentration and the increased hydrophobic contact angle.

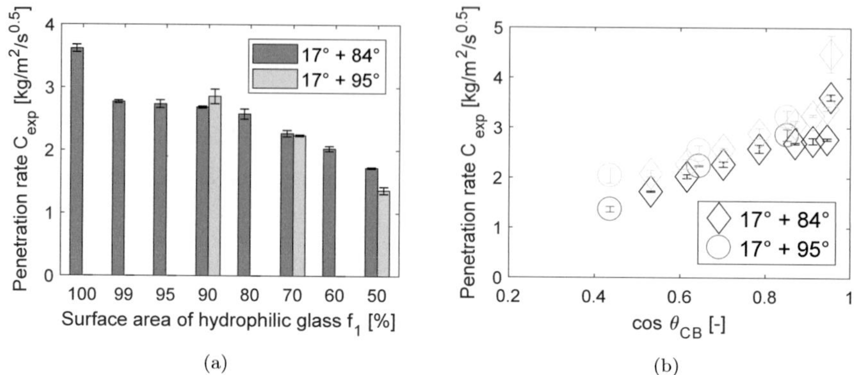

FIGURE 5.5: Experimental penetration rates C_{exp} over the surface area of hydrophilic glass (a) and over $\cos\theta_{CB}$ (b) for the fine mixtures of $17° + 84°$ and of $17° + 95°$.

In Figure 5.5 (b), the penetration rates of the coarse systems are shown in addition to the ones of the fine systems (light grey) in order to simplify the comparison. It can be seen that the capillary penetration into the coarse system occurs slightly faster over the whole contact angle range. This phenomenon was also observed by other authors (Dang-Vu and Hupka, 2009, Kiesvaara and Yliruusi, 1993, Staples and Shaffer, 2002). It can be explained by the presence of larger pores in the coarse system compared to the fine system which results in a higher rising velocity as long as the liquid level is not close to the equilibrium point. Besides the smaller pore radius, another reason for a slower rising velocity in the fine system is the increased frictional resistance to the liquid flow due to the higher total surface area of the glass beads.

5.2.3 Modelling of heterogeneous penetration

The modeling of penetration rates for the inert, heterogeneous systems was performed as described in section 4.1 and by using Eq. 4.4. Therefore, the Cassie-Baxter contact angle θ_{CB} was varied from $17.4°$ (pure hydrophilic) to $83.8°$ (pure hydrophobic). Table 3.5 provides the parameter of the water properties density, viscosity and surface tension. Furthermore, the porosity is taken from Figure 3.5. The effective pore radi of all combinations of hydrophilic and hydrophobic beads were calculated with Eq. 4.5 which involves the porosity, the capillary constants and the cross-sectional area of the Washburn cylinder of $7.85 \times 10^{-5}\,\mathrm{m}^2$. The values for the effective pore radii are presented in Table 5.1. Slight deviations between the radii can be observed, which can be explained by the randomness of the packing structure. But there is no influence of the content of hydrophobic surface on the calculated pore radius, as expected. The used equation for pore radius determination (Eq. 4.5) assumes a bundle of straight, parallel, cylindrical pores and does not consider the real pore shape, orientation and waviness. Nonetheless, by averaging the eight pore radii, a mean effective pore radius of $3.8\,\mu m$ for the inert powder beds is received and applied for the modelling. Dang-Vu and Hupka (2009) measured an effective pore radius of $3.9\,\mu m$ with the same equation for glass beads of 150 to $250\,\mu m$ size and toluene as ideal wetting liquid.

TABLE 5.1: Effective pore radius of different mixtures of hydrophilic and hydrophobic glass beads for coarse and fine system.

Surface fraction hydrophilic f_1 [%]	Surface fraction hydrophobic f_2 [%]	Coarse effective pore radius $r_{p,eff}$ [μm]	Fine effective pore radius $r_{p,eff}$ [μm]
100	0	3.91	2.89
99	1	3.55	2.61
95	5	3.80	2.61
90	10	3.80	2.96
80	20	3.80	2.76
70	30	3.60	2.83
60	40	3.86	2.55
50	50	3.82	2.51

Figure 5.6 presents the comparison of the experimental penetration rates for the two systems ($17°+84°$ and $17°+95°$) and the modelled penetration rates using the modified approach of Benavente et al. (2002). The black line corresponds to the modelling inserting the measured porosity of 0.40. Comparing experimental and modelled line for penetration using a porosity of 0.40, it is obvious that both are in the same size range, but the trend is different. While the experimental rate shows a significant jump in the graph with increasing hydrophobicity (decreasing cosine of θ_{CB}) the theoretical plot exhibits a smooth decrease. Without consideration of the hydrophilic penetration rate, the trend of the experimental and theoretical graphs looks similar. On the other hand, the experimental point for the pure hydrophilic sample is the only one matching the theoretical line. Benavente et al. (2002) introduced correction factors into their model equation in order to get a better fit of experimental and theoretical values. Roundness and tortuosity were implemented to describe the pore network more detailed. As in this thesis an effective pore radius was measured by applying the Washburn experiment with hexane geometric it is assumed that pore factors are already incorporated in the value $r_{p,eff}$.

One might try to explain this discrepancy by a deficient description of the solid-liquid interactions by the Cassie-Baxter contact angle. However, if the experimental penetration rate of $2.1\,kg\,m^{-2}\,s^{-0.5}$ of the 50:50-mixture (system $17°+84°$) is assumed to be located on the theoretical black line (Figure 5.6) the contact angle of this system would be 78.6°. In comparison to the contact angle of the pure hydrophobic material of 83.8° (at a contact line speed of $0.7 \times 10^{-3}\,\mathrm{mm\,s^{-1}}$), it is obvious that an angle of 78.6° for the whole system consisting of 50 % hydrophilic and 50 % hydrophobic glass beads is too high and, therefore, not realistic.

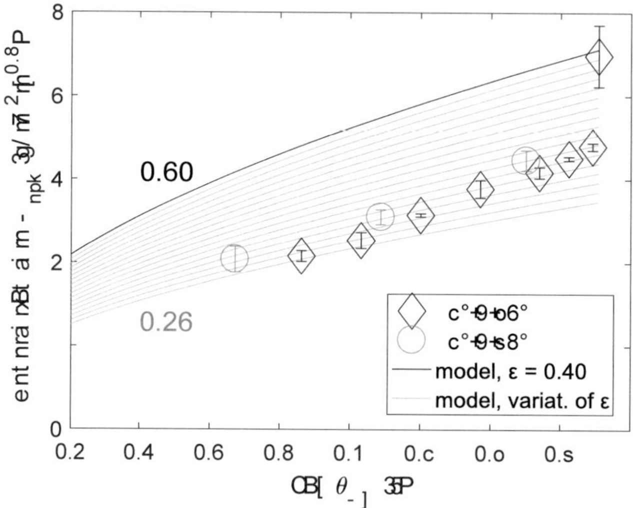

FIGURE 5.6: Comparison of experimental and modelled penetration rates for the coarse glass powder.

Thus, it is very likely that geometric factors are impacted by the addition of hydrophobic glass beads to the sample. The deviations between experimental and theoretical penetration rates can be explained by the change of pore volume due to the blockage of pores with the presence of hydrophobic walls. In Figure 5.6, the black line was modeled with a constant porosity of 0.40. But the blockage of pores leads to a decreasing effective porosity, as not all pores remain accessible, which has to be included in the model. Therefore, the light grey lines in Figure 5.6 represent the modelled penetration rates obtained by altering the effective porosity from 0.40 to 0.24. Within this porosity range the experimental points match the modeled penetration rates. It can be concluded that an addition of only 1 % of hydrophobic surface results in a reduction of effective porosity from 0.40 to approximately 0.30, while 50 % of the hydrophobic surface further decreases the porosity to 0.24. It follows that in a mixture of 50 % of hydrophilic glass beads and 50 % of hydrophobic ones, the accessible pore space is reduced by 40 %.

In Figure 5.7, the comparison of experimental and modelled penetration rates are shown for the fine glass powder. Resulting pore radii for capillary constant and porosity are also given in Table 5.1. As for the coarse system, it can be seen that the hydrophobicity does

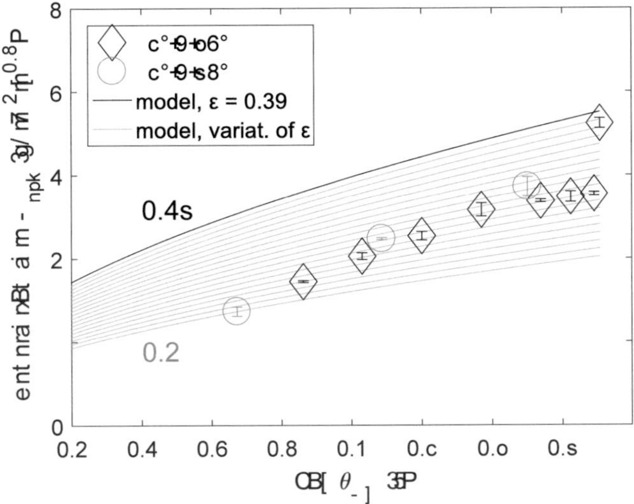

FIGURE 5.7: Comparison of experimental and modelled penetration rates for the fine glass powder.

not influence the calculated pore radii. Averaging the eight measured values an effective pore radius of 2.7 µm is obtained which is inserted in the model equation (Eq. 4.4). For the fine system, a porosity of 0.39 was determined and, thus, used for modelling the black line. The agreement of modelled black line and experimental data is insufficient as it was observed for the coarse system. Therefore, the porosity was varied again in order to include the pore blockage due to increasing hydrophobicity. An alteration of the porosity from 0.39 to 0.21 results in the coverage of all experimental penetration rates. The reduction from 0.39 to 0.21 changes the porosity by 46 %. In summary, it can be said that the experimental and theoretical penetration rates show a good agreement in the coarse and fine powder, if the effective porosity is adapted to account for blockage of pores with increasing hydrophobicity.

6

Capillary wetting into food systems

In the second chapter of results in this thesis, the hydrophilic glass material which behaves inert in contact with water was replaced by hydrophilic food materials which are soluble. Thus, the complexity of studying the liquid penetration into powdered systems is increased, since dissolution during liquid penetration leads to changing powder system properties as well as liquid properties.

6.1   Penetration into homogeneous systems

Firstly, the focus was placed on homogeneous food systems in order to investigate the solubility effect separately from the effect of contact angle. Therefore, the two dissaccarides, sucrose and lactose, were used to study mainly the influence of viscosity development. For determination of the influence of the pore network, penetration studies were performed with sodium chloride in different size fractions.

6.1.1   Influence of viscosity development on penetration rate

In this subsection, the water penetration into sucrose and lactose powder was investigated experimentally. Therefore, the Washburn setup described in section 2.5 was used. Besides experimental investigations, a model was developed which can deal with liquid penetration driven by capillary forces and dissolution at the same time. The model is described in detail in section 4.2. The parameters which are needed for the model were either taken from literature or were determined with experiments. The capillary constant and the porosity of sucrose and lactose powder are presented in subsection 3.3.2. Both values are used to calculate the effective pore radius $r_{p,eff}$ (Eq.4.5) which is needed for the modelling. The size of the particles influencing the dissolution process in the model is estimated by the $d_{50,3}$-value which is also given in subsection 3.3.2. Furthermore, the contact angles at the most dynamic point corresponding to the highest contact line speed were inserted into the model equations, since at the beginning of the capillary rise a fast penetration velocity can be observed. For sucrose, a contact angle of 31.3° at a contact line speed of $10\,\mathrm{mm\,s^{-1}}$ was measured and for lactose, a contact angle of 24.4° at a contact line speed of $4.9\,\mathrm{mm\,s^{-1}}$.

The load depending solution properties for sucrose and lactose such as viscosity, density and surface tension, were taken from litertaure and are presented in Figures 6.1 (a)-(c). The graphs were only plotted within the load range which was covered with experimental data in literature.

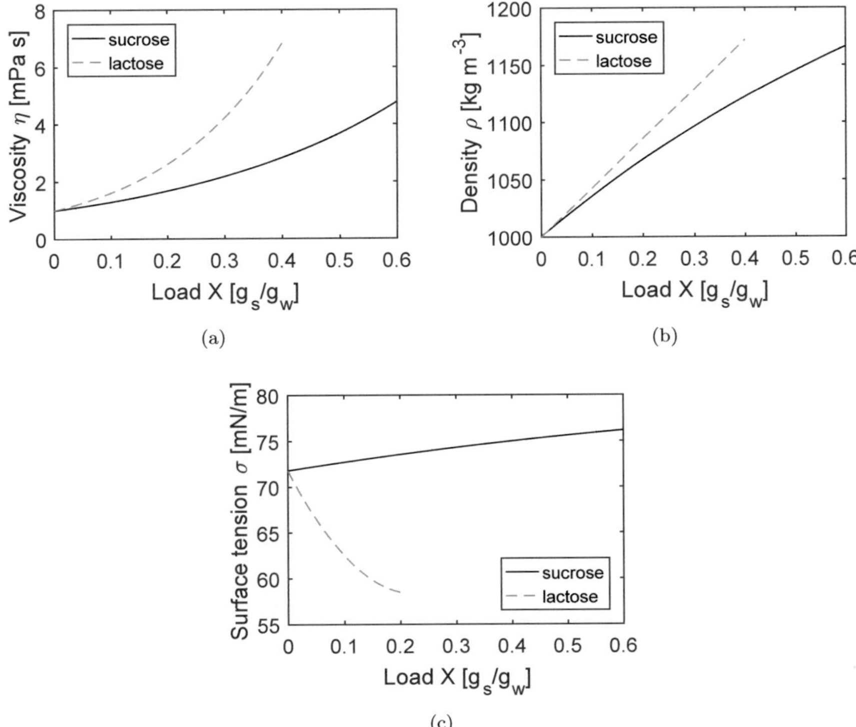

FIGURE 6.1: Viscosity (a), density (b) and surface tension (c) as a function of load for a sucrose and a lactose solution. Values were taken from literature. Sucrose: viscosity and density from Asadi (2007) and surface tension from Honig (1953). Lactose: viscosity and density from Walstra (1999) and surface tension from Adhikari et al. (2007).

For the viscosity and density, it can be seen that the loading of water with lactose leading to a higher increase compared to sucrose. The viscosity rises by 2.8 times and 6.8 times for sucrose and lactose, respectively, within a load range from pure water to $0.4\,\mathrm{g_s/g_w}$. In the same load range, the density increase of 13 % and 17 % for sucrose and lactose is relatively lower. A different trend is observable for surface tension. The addition of sucrose to water $(0.2\,\mathrm{g_s/g_w})$ leads to a small increase by 2 % in surface tension, whereas lactose solution at the same load decreases the surface tension of pure water by 19 %. Thus, it can be concluded that the lactose powder includes surface active components which can be traces of milk proteins. Comparing the three liquid properties for sucrose and lactose solutions, it is obvious that dissolution has the major influence on viscosity. Therefore, viscosity is the main factor which has to be considered during liquid penetration into soluble systems.

Further modelling parameters are summarized in Table 6.1. The saturation densities of sucrose and lactose solutions are used in the mass transfer equation to calculate the driving gradient. It is obvious that the saturation density of sucrose is higher than the one of lactose. This is due to the fact that sucrose is better soluble in water leading to a higher saturation concentration and, hence, to a higher saturation density (compare with section 3.1). Load depending and, thus, time depending diffusion coefficients are determined in the model using the Stokes-Einstein equation (Eq. 4.12) which requires the hydrodynamic radius of the sugar molecules in water. All parameters are given at a temperature of 20 °C.

TABLE 6.1: Saturation density and hydrodyamic radius of sucrose and lactose at 20 °C for modelling (Mathlouthi and Reiser, 1995, McDonald and Turcotte, 1948, Mohos, 2017, Schultz and Solomon, 1961).

Parameter	Sucrose	Lactose
Saturation density $[kg/m^3]$	1332.7	1074.2
Hydrodynamic radius $[nm]$	0.49	0.54

All parameters calculated by the model for the penetration into sucrose and lactose were evaluated within a range of 0 to 6 s. Since the utilized Sherwood correlation is only applicable in the range of Reynolds numbers from 0.1 to 10^4, the model loses its validity after 5.8 s in case of sucrose. The Reynolds number stayed within the limits for modelling penetration into lactose. Another factor which has to be considered are the limits of the Schmidt number from 0.6 to 10^4 for using the Sherwood correlation. While for lactose the Schmidt number is within the boundaries during the analyzed 6 s, the Schmidt number gets above a value of 10^4 already after 1.6 s for sucrose. However, the Sherwood correlation was used also in the case of sucrose, as there is no other correlation available. Furthermore, no suspicious trend was observed after the 1.6 s for the different modelled parameters.

Figure 6.2 provides the solution of Eq. (4.6) and presents the load of the penetration flow over time. A fast increase of sucrose load from 0 to $0.49\,\mathrm{g_s/g_w}$ within 6 s can be observed, while the lactose load reaches only a value of $0.09\,\mathrm{g_s/g_w}$ after 6 s. This deviation is explainable by the different saturation densities of sucrose and lactose. A higher saturation density in the case of sucrose leads to a larger gradient which is the driving force for dissolution. The model assumes an ideal mixing of the dissolved component in water which results in a constant concentration over the whole solution. With an increasing load in the dissaccaride solutions, the liquid properties change.

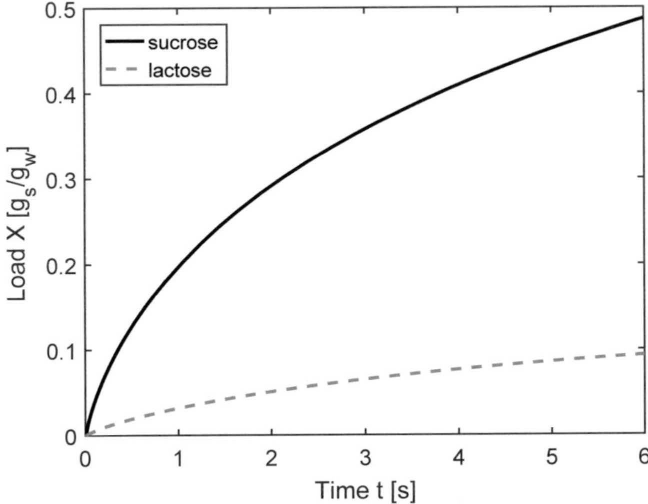

FIGURE 6.2: Load of sucrose and lactose solutions as a function of time predicted by the model.

The alteration of the three liquid properties viscosity, density and surface tension as a function of time predicted by the model is presented in Figures 6.3 (a)-(c). The viscosity plot (a) shows a large increase for the sucrose solution up to $4.0\,\mathrm{mPa\,s}$ within the analyzed time interval, whereas the lactose solution reaches only a viscosity of $1.3\,\mathrm{mPa\,s}$. A similar trend can be observed in the density plot (b). The sucrose solution increases its density by 14 %, while the increase of the density of the lactose is significantly lower with 4 % . Thus, it can be concluded that during a relatively short time period of 6 s both viscosity and density increase significantly due to dissolution of the dissaccarides. However, the increase for the sucrose solution is clearly higher for both liquid parameters due to the faster dissolution of sucrose in water. The surface tension as a function of time calculated by the model behaves differently. A small increase of 6 % within 6 s was identified for the sucrose solution. In case of the lactose solution, the surface tension decreases by 13 % due to the presence of surface active components. Comparing the relative alteration of the three liquid properties within the observed time interval, the viscosity has with increases of 300 % and 30 % for sucrose and lactose, respectively, the major impact on the liquid penetration in our model. Both the mass transfer and the capillary rise are reduced with increasing viscosity. The capillary rise dh/dt is directly influenced by the viscosity, whereas a reduced velocity and an increased viscosity decrease the Reynolds number and the diffusion coefficient. On the other hand, the Schmidt number is increased by an increased viscosity and a decreased diffusion coefficient. But since the decrease of the Reynolds number is relatively higher than the increase of the Schmidt number, the Sherwood number and, thus, the mass transfer coefficient are decreased leading to the reduction of mass transfer dM_s/dt.

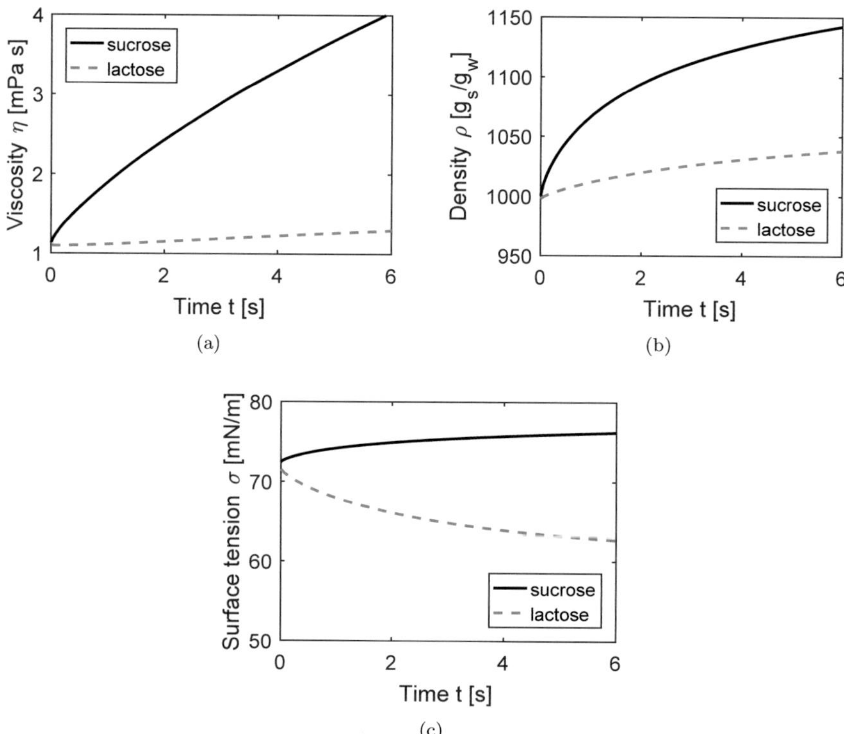

FIGURE 6.3: Viscosity (a), density (b) and surface tension (c) as a function of time for a sucrose and a lactose solution predicted by the model.

The model includes the change of liquid properties due to dissolution, but neglects the decreasing particle size and the resulting increase of pore volume. Therefore, a comparison is drawn in order to see the effect of particle volume and of viscosity which is the mainly influencing liquid property. Since sucrose is the better soluble material, the comparison is carried out using the sucrose case. After 6 s, the total amount of dissolved sucrose is calculated by the load ($0.49\,\mathrm{g_s/g_w}$) and the penetrated water mass ($0.58\,\mathrm{g_w}$) and results in $0.28\,\mathrm{g_s}$. Knowing the porosity in the sample holder 0.42, the density of water ($998\,\mathrm{kg/m^3}$) and the density of sucrose ($1553\,\mathrm{kg/m^3}$), a sucrose mass of $1.23\,\mathrm{g_s}$ which is in contact with water after $6\,\mathrm{s}$ can be determined. Thus, $23\,\%$ of the involved sucrose volume is dissolved within the observed time interval which can be counted as additional pore volume. On the other hand, the dissolution of $0.28\,\mathrm{g_s}$ sucrose in $0.58\,\mathrm{g_w}$ water results in a viscosity increase of $300\,\%$. Therefore, the neglection of the change in particle size seems to be tolerable when investigating the penetration during the first seconds.

In order to evaluate the new model, experimental data is compared to the data predicted by the model. The experimentally measured squared mass of water as a function of time for the sucrose and the lactose samples and the corresponding modelled graphs are presented in Figure 6.4.

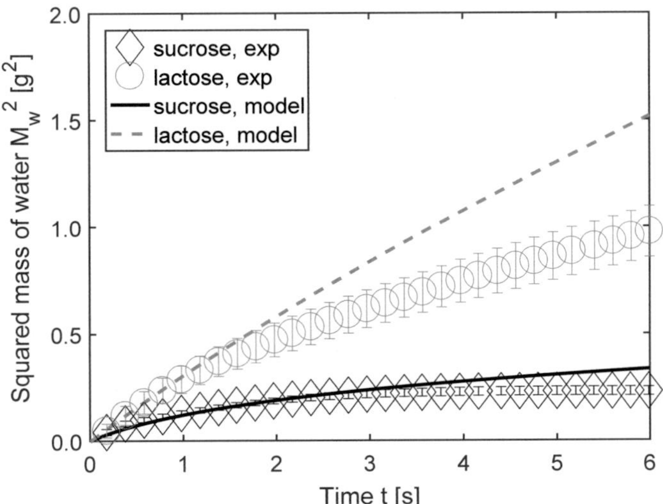

FIGURE 6.4: Squared mass of water as function of time during penetration into sucrose and lactose powder: experimental data in comparison to data predicted by the model.

Due to less dissolution of lactose in water and a lower viscosity increase compared to sucrose, the increase of the squared mass-plot is significantly higher for lactose. This trend is also predicted by the model. Comparing the experimental and modelled graphs, a similar trend can be observed within the first few seconds. After about $1.5\,\mathrm{s}$ for the lactose graphs and after $3.5\,\mathrm{s}$ for the sucrose graphs, the experimental water penetration rate slows down, while the model starts to overestimate the penetration. In the experiment, a beginning collapse of the powder beds can be observed resulting in the partial blockage of pores.

Another expression of the squared mass data is given by the penetration rates as they were already introduced in section 5.2. For determination of the experimental penetration rate and the penetration rate of the model with dissolution, the slope of the linear increase of the squared mass-time-plot was divided by the cross-sectional area of the cylinder. A coefficient of determination R^2 of at least 0.98 is set for selection of the linear part and its slope. The time interval was chosen from $0\,\mathrm{s}$ to at least $2.3\,\mathrm{s}$ depending on the length of the linear part. Furthermore, the approach of Benavente et al. (2002) was also applied to determine an inert penetration rate. Thus, three penetration rates are compared for each dissaccaride.

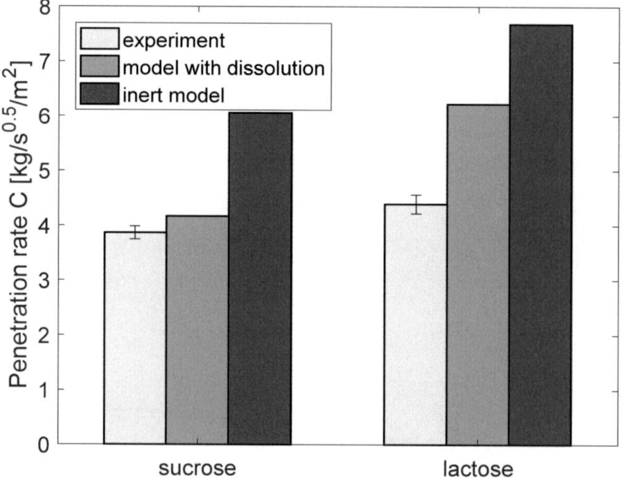

FIGURE 6.5: Penetration rates into sucrose and lactose powder determined by experiments, with the developed model for dissolution and with the inert model.

The penetration rates which are determined by the experiment are $3.9\,\mathrm{kg\,m^{-2}\,s^{-0.5}}$ for sucrose and $4.4\,\mathrm{kg\,m^{-2}\,s^{-0.5}}$ for lactose. When comparing experimental and modelled penetration rates it is obvious that water penetration into a soluble food powder such as sucrose cannot be described by the model for inert systems which values overestimate the penetration by $56\,\%$ and $75\,\%$ for sucrose and lactose, respectively. The inert model overestimates the real penetration rates due to several reasons, which are all related to the dissolution during the wetting process. Consequences of dissolution are the decrease of particle size, the resulting increase of pore volume and the decrease of surface area of particles and the change of liquid properties due to the load. Therefore, the developed model dealing with continuously changing liquid properties due to dissolution provides already a better prediction of the penetration behavior. In case of sucrose the deviation for the dissolution model is only 8 %, whereas the penetration into lactose is overstimated by 40 %. In summary, it is apparent that the new model accounting for dissolution shows a fair agreement with experimental data for the sucrose system which is a fast dissolving system with a strong increase of viscosity. Since viscosity increase is the main decelerating factor for liquid penetration into the sucrose powder which is considered in the model, model and experiment are controlled by the same factor. In case of the lactose system, the viscosity increase is lower and, thus, not the main decelerationg factor. It seems that partial collapse of the lactose bed and the resulting partial pore blockage plays a major role during penetration which is not considered in the model.

6.1.2 Influence of particle size on penetration rate

Besides the investigation of penetration into carbohydrate powders which are known for viscosity developing during dissolution, secondly, it was focused on the penetration into different size fractions of sodium chloride in order to study the influence of particle and pore size. An experimental analysis was performed as well as the modelling of penetration and dissolution. Sodium chloride is also well soluble in water with a saturation concentration of $0.36\,\mathrm{g_s/g_w}$. But compared to the carbohydrate powders, the dissolution of sodium chloride in water does not lead to a sharp increase of viscosity as it can be seen in Figure 6.6 (a). Within an increase of load from 0 to $0.5\,\mathrm{g_s/g_w}$, the viscosity rises by 1.9 times. The increase of density from 0 to $0.5\,\mathrm{g_s/g_w}$ by 19 % is in the same range as for sucrose and lactose solutions (Figure 6.6 (b)). The third liquid parameter which is affected by dissolution is the surface tension increasing by 14 % as it is shown in Figure 6.6 (c). Compared to the dissaccarids, the surface tension trend for sodium chloride is similar to the one of sucrose, but the increase is slightly higher. The data for liquid properties were taken from literature (Harkins and McLaughlin, 1925, Kestin et al., 1981, Pitzer et al., 1984) and expressed by fitted equations. For predicting the penetration into sodium chloride powder, those fitted equations were implemented in the model.

Furthermore, the saturation density and the hydrodynamic radius are required for the modelling and were also taken from literature (Hussain et al., 2006, Thurmond et al., 1984). The saturation density is used to describe the driving gradient for the mass transfer and the hydrodynamic radius is required for the Stokes-Einstein equation to express the diffusion coefficient. Both parameters are presented in Table 6.2. The saturation density for sodium chloride is in between the densities for lactose and sucrose and, thus, the solubility as well. Since sodium chloride dissociates into sodium and

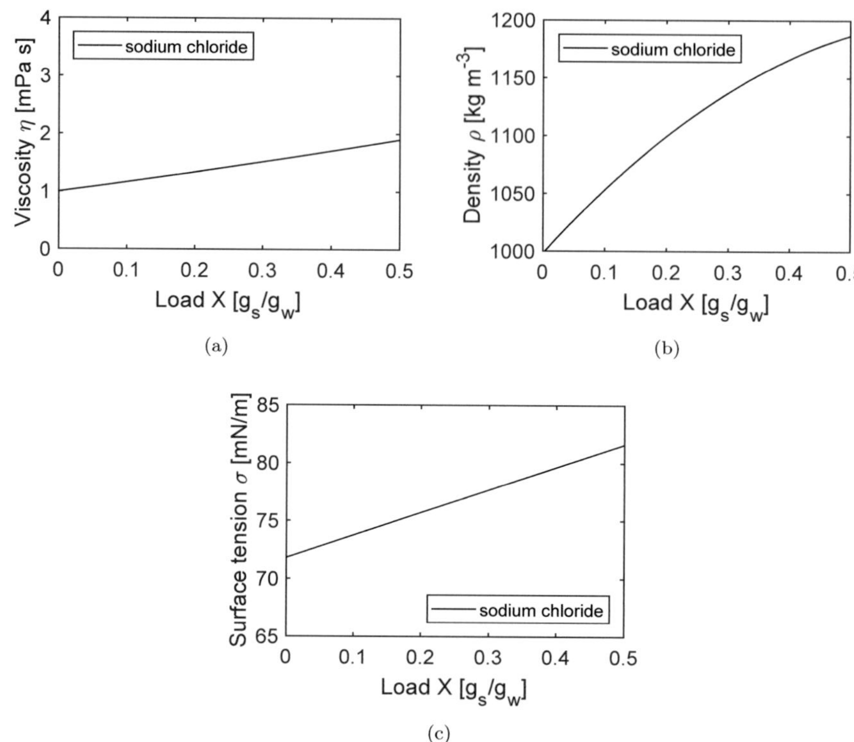

FIGURE 6.6: Viscosity (a), density (b) and surface tension (c) as a function of load for sodium chloride solution. Values were taken from literature: viscosity from Kestin et al. (1981), density from Pitzer et al. (1984) and surface tension from Harkins and McLaughlin (1925).

chloride ions when dissolving in water, an averaged hydrodynamic radius of both ions was calculated and used in the model.

TABLE 6.2: Saturation density and hydrodyamic radius of sodium chloride at 20 °C for modelling (Hussain et al., 2006, Thurmond et al., 1984).

Saturation density $[kg/m^3]$	1200.9
Hydrodynamic radius $[nm]$	0.15

The geometric parameters which need to be inserted into the model were determined by experiments for each size fraction of sodium chloride. The measured particle size distributions provide the $d_{50,3}$-values which is used as equivalent particle diameter for describing the dissolution process. For the four size fractions, the values range from 205 µm to 674 µm. Moreover, the porosities and capillary constants were determined as it is presented in subsection 3.3.2 and implemented in the model. Porosities and capillary constants are further used to calculate the resulting effective pore radii. The values are 5.4 µm, 5.6 µm, 6.1 µm and 4.8 µm for s_1, s_2, s_3 and s_4, respectively. The lowest effective pore radius for the largest particle size does not mean that the pore size within this system is the finest in reality. But it is very likely that macrovoids are present in the largest powder leading to pore blockages and to a drecreased capillary constant and, thus, to a lower effective pore radius.

All parameters calculated by the model for the penetration into the four fractions of sodium chloride were evaluated within a range of 0 to 10 s. Both, the Reynolds number and the Schmidt number stay within the validity limits from 0.1 to 10^4 and 0.6 to 10^4, respectively, for using the Sherwood correlation for flow through a packed bed.

Figure 6.7 provides the solution of Eq. (4.6) and shows the load of the penetration flow which is calculated from the mass of dissolved sodium chloride per mass of water as function of time for four different size fractions. The model assumes an ideal mixing of the dissolved component in water which results in a constant concentration over the whole solution.

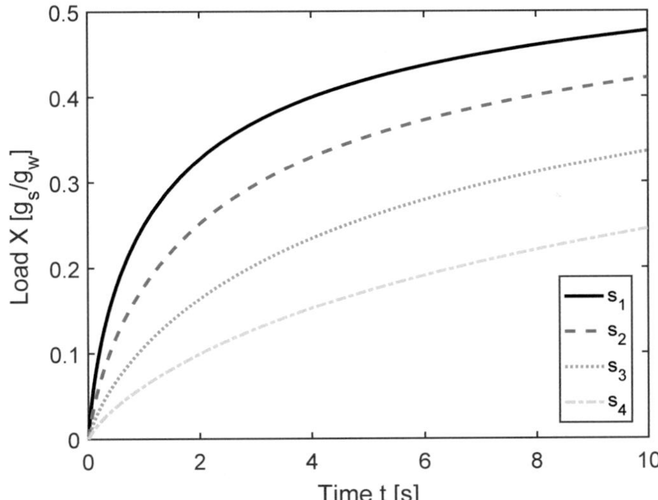

FIGURE 6.7: Load of sodium chloride solutions for four different size fractions as a function of time predicted by the model.

For all four samples, a significant increase of sodium chloride load in water can be observed with a decreasing trend with increasing particle size. While the load of solution in the s_1-sample rises to a value of 0.48 $\mathrm{g_s/g_w}$, the solution in the coarsest sample reaches a load of 0.25 $\mathrm{g_s/g_w}$ after 10 s. This trend is linked to the increased surface area with decreasing particle diameters. Therefore, in the finer samples, there is more contact area between water and sodium chloride resulting in an increased mass transfer. The loading of water with sodium chloride lead to the change of liquid properties. Due to the direct proportionality of load and liquid properties, the alteration of viscosity and density follows the same increasing trend with time. In Figure 6.8, the viscosity and density as a function of time are presented. The surface tension plot is not shown since the maximum alteration is with 13 % the lowest of the three liquid properties. Concerning the particle size fractions, the same trend as for the load can be seen. The larger the particles, the smaller is the increase of viscosity and density within the observed time intervall. For viscosity, an increase of maximum 85 % is achieved for the s_1-solution. The density rises by 19 % for the s_1-solution after 10 s. In comparison to the carbohydrate samples, the viscosity increase for sodium chloride is in between of sucrose and lactose. In case of density, for sodium chloride the increase is higher than for sucrose and lactose. The trend for the change of surface tension is similar to the one of sucrose (compare with Figure A.1 in the Appendix A). Thus, investigating the penetration into sodium chloride powder, the liquid properties are decisive factors influencing the rise.

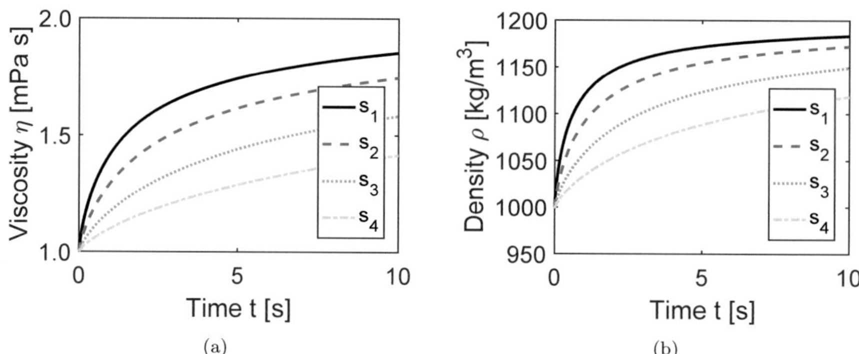

FIGURE 6.8: Viscosity (a) und density (b) as a function of time for sodium chloride solutions for four different size fractions predicted by the model.

Since the change of geometric parameters of the powder bed due to dissolution is neglected in the model, the effect of dissolution on the change of particle volume for sodium chloride is calculated and evaluated. The sample s_2 is chosen for this calculation because of its similar particle size to sucrose and lactose. After 10 s, the total amount of dissolved sodium chloride s_2 is calculated by the load (0.42 g$_\text{s}$/g$_\text{w}$) and the penetrated water mass (1.0 g$_\text{w}$) and results in 0.42 g$_\text{s}$. Knowing the porosity in the sample holder 0.50, the density of water (998 kg/m^3) and the density of sodium chloride (2210 kg/m^3), a sodium chloride mass of 2.2 g$_\text{s}$ is in contact with water after 10 s. Thus, merely 19 % of the involved sucrose volume is dissolved within the observed time interval. It can be concluded that there is an effect of dissolution on particle volume during wetting.

But since the alteration of viscosity is higher within the time intervall, the neglection of particle volume seems to be tolerable in a first step.

For evaluation of the model and to proof the applicability for sodium chloride, experimental and modelled data are compared. The experimentally measured squared mass of water as a function of time for the four sodium chloride samples and the corresponding modelled graphs are presented in Figure 6.9 (a) and (b).

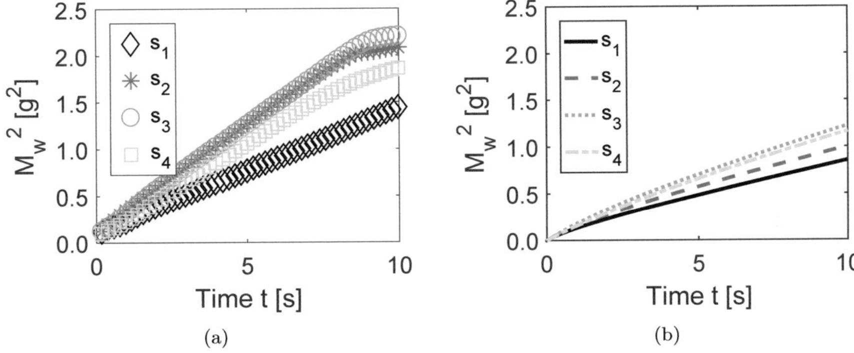

FIGURE 6.9: Squared mass of water as function of time during penetration into four different size fractions of sodium chloride powder determined by experiments (a) and predicted by the model (b).

First of all, it is obvious that the trend has changed compared to the load and liquid properties graphs. With the increasing particle size from s_1 to s_2 in Figure 6.9 (a), the slope increases, while a further increase of particle size to s_3 does not lead to a change of slope. But a further increase of particle size to s_4 results in a sudden decrease of the slope of the squared mass of water versus time plot. In Figure 6.9 (b), the behaviour of the modelled graphs looks slightly different. From s_1 to s_3, an increasing slope can be observed, while the slope of s_4 only marginally decreases. The reducing trend for s_4 was also observed for the capillary constant (subsection 3.3.2). Hence, it can be concluded that the conditions of the pore network lead to this decrease. For the smallest size fraction s_1, the capillary constant showed a different behaviour. While the capillary constant of s_1 is very close to s_2 and s_3, the slope of the M^2-t-plot of s_1 shows the largest deviation to the other three plots in the experiment as well as in the model. It seems that the smallest size fraction which results in the highest viscosity and density development during contact with water is more affected by dissolution than the other three samples. Comparing the experiment and the model, the slopes of the graphs in the experiment are continously higher than the slopes predicted by the model. Since the model does not predict the influence of dissolution in terms of changing pore network properties, the model underestimates the reality. The alterating liquid properties in the model only reduce the penetration peformance while the decreases of particle size and consequent increases of pore size would enhance the penetration velocity and, thus, the slope of the M^2-t-plot. A comparison of each experimental and each modelled graph (M^2-t-plot) is given in Figures A.2 (a) - (d) in the Appendix A.

Furthermore, the experimental and modelled M^2-t-plots were converted into penetration rates as it was described for sucrose and lactose in subsection 6.1.1 and are presented in Figure 6.10. Additionally, the inert penetration rates calculated by the approach of Benavente et al. (2002) are shown. It is obvious that the penetration rates predicted by the inert model are closer to the experimental rates than the ones predicted by the new model. The rates determined by the model considering dissolution are significantly lower. This effect is caused by the inclusion of changing liquid properties during penetration which reduce the penetration performance. On the other hand, the positive influence on the pore network by pore size enlargement due to dissolution is neglected. Therefore, by neglection of both, liquid and pore network properties, the inert model shows a better fit.

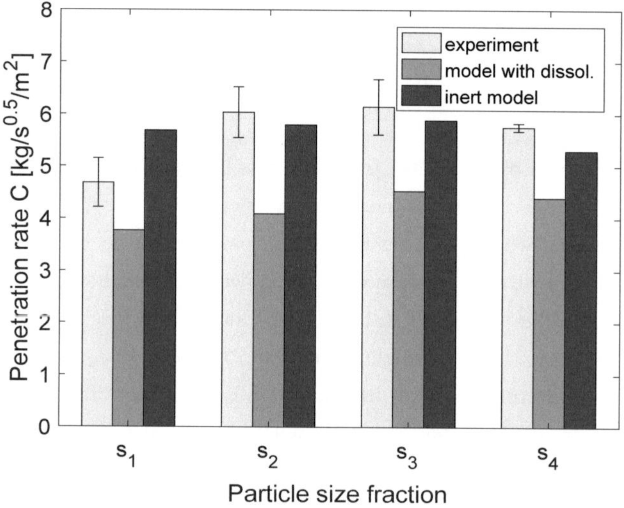

FIGURE 6.10: Penetration rates into four different size fractions of sodium chloride powder determined by experiments, with the developed model for dissolution and with the inert model.

As a final conclusion of this subsection, it could be shown that in case of sodium chloride powder which increases the liquid properties during dissolution, but does not increase the viscosity in such a intense way like sucrose, the inert model was in better agreement with the experiment than the new model. For improving the new model in order to give a better fit, besides the consideration of changing liquid properties the effect of dissolution on particle size and, thus, on pore size should be included.

6.2 Penetration into heterogeneous systems

In this section, complexity is further increaseed by studying the combination of two influencing factors during wetting. Hydrophilic and soluble food powders were mixed with hydrophobic, inert components in order to investigate the effect of dissolution and the effect of hydrophobicity on the the penetration of liquid into the powder mixture. Therefore, firstly, sucrose powder which is known as viscosity developing component during dissolution was mixed with hydrophobic glass powder. Secondly, sodium chloride powder as soluble component which does not change the liquid properties significantly during dissolution within the observed time intervall was used as hydrophilic component and combined with shellac coated glass beads.

6.2.1 Effect of viscosity and heterogeneity on liquid penetration

Experimental penetration of liquid into a mixed system of hydrophilic, soluble sucrose and hydrophobic, inert glass powder is studied in this subsection. Furthermore, the developed model is applied to predict the penentration behaviour. Experimental and modelled data are compared and evaluated. Four powder combinations with different surface area proportions reaching from 100 % sucrose to 50 % sucrose and 50 % glass beads were used. A summary about the mixtures and their resulting contact angles which are required for modelling is provided in Table 3.8. By means of the volume specific surface areas of sucrose (from Table 3.2) and glass (from Table 3.1) and the material densities of sucrose (from subsection 3.3.2) and glass (from subsection 3.3.1), the fractions of surface area can be converted into the fractions of mass, which were used for preparation of the samples. The powder properties such as the porosities and the capillary constants of all combinations are presented in subsection 3.3.3. By means of the capillary constants and the porosities, effective pore radii of the packings were determined (Eq. 4.5) in order to insert the values into the model which are presented in Table B.1 (Appendix B).

Figure 6.11 presents the sucrose load of the penetration flow as a function of time for the four different mixtures of sucrose and glass beads. A fast increase of sucrose load from 0 to $0.49\,\mathrm{g_s/g_w}$ within 6 s can be observed for the 100 % sucrose sample. With the addition of 5 %, 20 % and 50 % of hydrophobic glass surface, the sucrose load of the solution after 6 s is reduced to $0.44\,\mathrm{g_s/g_w}$, $0.40\,\mathrm{g_s/g_w}$ and $0.31\,\mathrm{g_s/g_w}$, respectively. This decrease of dissolved sucrose in the liquid flow is explainable by the reduced contact area between liquid and sucrose due to the presence of glass surface. Since the model assumes an ideal mixing of the dissolved component in water, the sucrose load is constantly distributed over the whole solution. With an increasing sucrose load of the liquid within the powder, the liquid properties change.

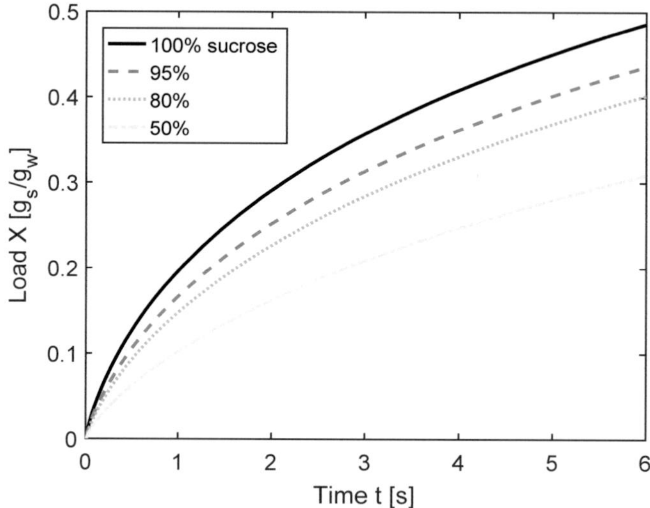

FIGURE 6.11: Load of sucrose solution for penetration into systems consisting of sucrose and hydrophobic glass as a function of time predicted by the model.

The alteration of viscosity and density within a time interval of 6 s predicted by the model is presented in Figures 6.12 (a) and (b). Since the change of surface tension is only marginal within 6 s, the focus is drawn on viscosity and density. The viscosity plot (a) shows a large increase for the sucrose solution up to 4.0 mPa s within the analyzed time interval, whereas the density (b) rises to 1142 kg m^{-3} for the 100 % sucrose sample. With an increasing amount of hydrophobic surface in the sample, the viscosity as well as the density decreases as it was already seen for the load-time-plot. Thus, it can be concluded that during a relatively short time period of 6 s, viscosity and density increase significantly due to dissolution of the sucrose, while an increase of hydrophobic glass surface counteracts this trend.

In order to evaluate the influence of both, solubility and hydrophobic surface on the liquid penetration into the sample, Figure 6.13 provides the squared mass of water-time-plot for the four different cases determined with experiments (a) and predicted by the model (b). It is immediately apparent that the trend has changed in these diagrams. For a better understanding of this behaviour, it has to be distinguished between two main effects, which are linked to the wetting of hydrophobic surfaces and the wetting of

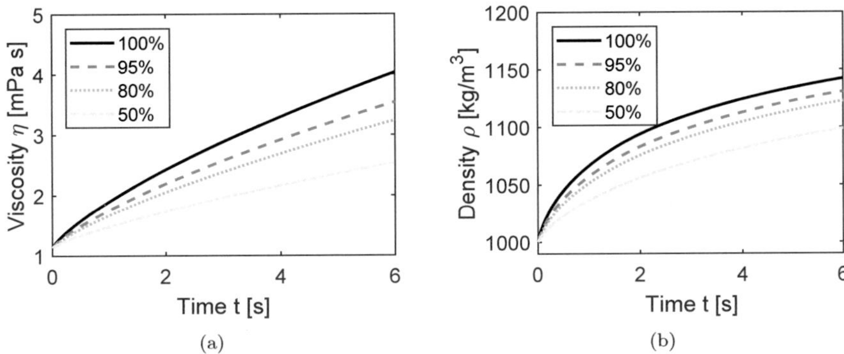

FIGURE 6.12: Viscosity (a) und density (b) of sucrose solution for penetration into systems consisting of sucrose and hydrophobic glass as a function of time predicted by the model.

soluble material. An increasing amount of hydrophobic surface decreases the capillary penetration into a powder bed. This effect was described in a study of Kammerhofer et al. (2018a). Secondly, the contact of water and sucrose results in the dissolution of sucrose leading to an increasing viscosity of the liquid. The increased viscosity has also a decelerating influence on the water penetration (Kammerhofer et al., 2018b). Thus, we have to consider two factors slowing down the capillary wetting, but both in a different way.

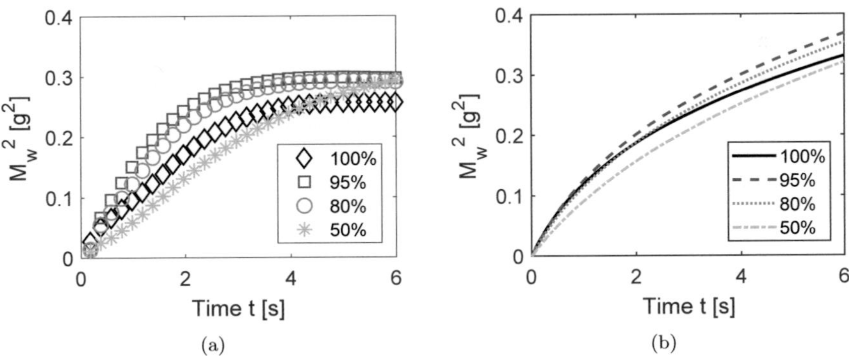

FIGURE 6.13: Squared mass of water during penetration into systems consisting of sucrose and hydrophobic glass as a function of time determined by experiments (a) and predicted by the model (b).

In case of the experimental diagram, an increasing trend for the slope from 100 % to 95 % and 80 % can be observed, while the addition of 50 % of hydrophobic glass reduces the slope. The modelled plots in Figure 6.13 (b) show a similar behaviour. During the first seconds the slope of the 100 % sample is very close to the ones of 95 % and 80 %, while the slope of the 50 % is significantly lower. During the initial time period, the hydrophobic surface has the major reducing impact on the penetration velocity. After about 2 s, the slope of the 100 % sucrose-curve starts to decrease significantly faster compared to the samples 95 % and 80 % and approximates the slope of the 50 %-curve. It is obvious that during the second period starting at approximate 2 s, the impact of the viscosity increase due to dissolution gains importance. The fast increase of viscosity in the 100 % sample results in a decelerating effect on the rise, while the addition of 5 % and 20 % of hydrophobic surface to the samples seems to have a stabilizing impact and increases liquid flow. This increasing effect is very likely due to the change of properties of the pore network which occur by adding spherical glass beads to the sucrose powder. Furthermore, it seems that the increasing contact angle due to hydrophobic glass (in small or intermediate quantities) in the system is balanced by the reduced dissolution and the consequent reduced viscosity increase, since there is less contact area between sucrose and water. In the 50:50-mixture, a different trend is observed. Even if the contact area between sucrose and water is further decreased, the high amount of hydrophobic surface area in this sample leads to a decrease in penetration flow. Hence, we distinguish between two concentration ranges within the observed compositions. Firstly, a concentration range of hydrophobic surface area could be determined where the effect of hydrophobic surface on the contact angle is balanced by the reduced dissolution at small and intermediate quantities. Secondly, at high quantities a decelerating effect of hydrophobic surface during the capillary wetting of sucrose was observed due to the increase of contact angle.

For a better comparison of each experimental curve with standard deviation and the modelled plot, Figure 6.14 provides four diagrams for the different mixtures. Except for the pure sucrose sample, the graphs predicted by the model are almost within the standard deviation of the experimental data. However, it seems that the model overestimates the reality at the end of the time interval. This can be explained by the neglection of the partial pore blockage when due to dissolution the pore volume gets larger and collapses which is not considered in the model. In case of the pure sucrose sample, the model provides an overestimation over the whole time interval. This is very likely due to the highest dissolution rate in the 100 % sample which leads also to the largest reduction in particle volume and, thus, to a partial collapse of pores.

Furthermore, a comparison of experimental and modelled penetration rates is presented in Figure 6.15. They are obtained by the slope of the linear increase of the plots in Figure 6.13. The determination is described in detail in subsection 6.1.1. The trends discussed above can be also observed for the penetration rates. The addition of hydrophobic surface area of 5 % and 20 % leads to an increase of the experimental penetration rates from $3.87 \, \mathrm{kg \, m^{-2} \, s^{-0.5}}$ (100 %) to $4.12 \, \mathrm{kg \, m^{-2} \, s^{-0.5}}$ (95 %) and $4.06 \, \mathrm{kg \, m^{-2} \, s^{-0.5}}$ (80 %).

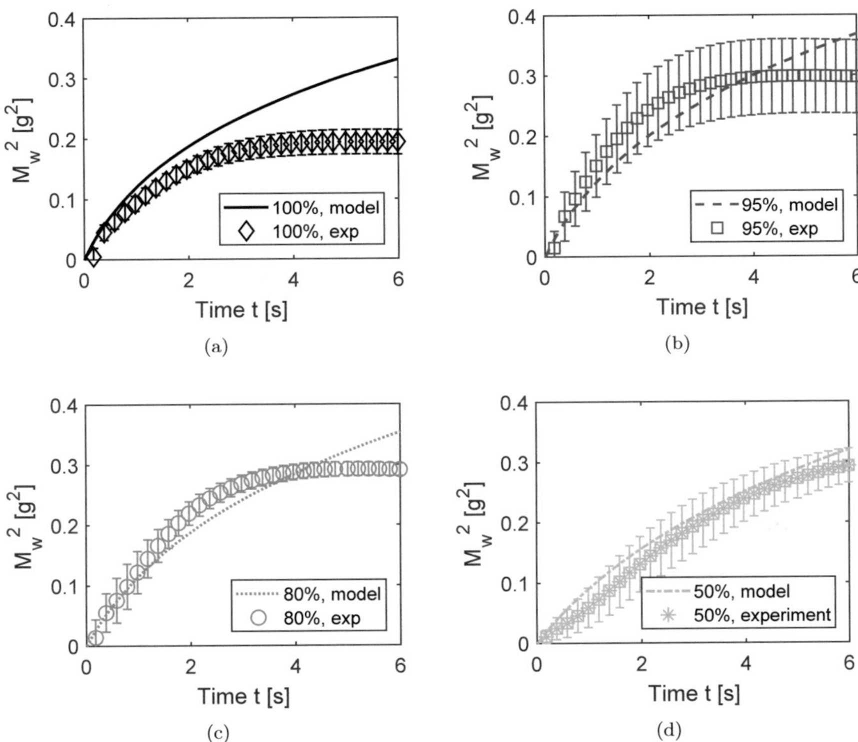

FIGURE 6.14: Squared mass of water as function of time during penetration into systems consisting of sucrose and hydrophobic glass. Comparison of experiment and model for sample containing 100 % sucrose surface (a), 95 % sucrose surface (b), 80 % sucrose surface (c) and 50 % sucrose surface (d).

Due to the high standard deviations the penetration rates of 100 %, 95 % and 80 % do not differ significantly, even if an increasing trend is observable. A further addition of hydrophobic surface area of 50 % results in a significant decrease in penetration rate to $3.30\,\mathrm{kg\,m^{-2}\,s^{-0.5}}$. A fair agreement of experimental and modelled penetration rates can be observed for the 95 %, 80 % and 50 % sample, since the modelled rates are within the standard deviation of the experimental ones. In case of the pure sucrose sample, the model overstimates the experiment which is already explained above. However, in summary, the developed and adapted model provides a good opportunity to predict the penetration behavior into heterogeneous and soluble particle systems.

Comparing all samples the mixture containing 5 % hydrophobic surface provides the best result in terms of fast liquid penetration for the experimental investigation as well as for the model. The increase of the apparent mixed contact angle from 31.3° to 35.2° with the addition of 5 % of hydrophobic surface appears to be balanced by the slower increase

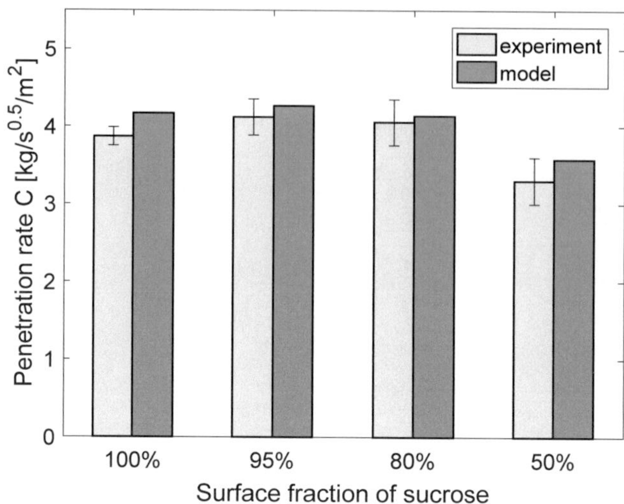

FIGURE 6.15: Penetration rates into four sucrose-glass-mixtures determined by experiments and with the developed model for dissolution.

of viscosity. Furthermore, the slightly larger capillary constant of the 95 %-sample offers a better condition of the pore network for liquid penetration compared to the 100 %-sample. While the 80 %-sample have an even larger capillary constant, the increased hydrophobicity expressed by a contact angle of 45.2° counteracts a further enhancement for the penetration flow.

6.2.2 Effect of dissolution and heterogeneity on liquid penetration

In this subsection, the liquid penetration into mixtures of sodium chloride powder (s_2) and shellac coated glass beads was studied. Therefore, the experimental penetration was measured as described in section 2.5. The samples were mixed in order to get surface proportions of 100:0, 95:5, 80:20 and 50:50. Additonally, the developed model for liquid penetration into heterogeneous and soluble systems was used for comparison with experiments and evaluation. The contact angles which are inserted in the model were taken from Table 3.9. Furthermore, the properties of the pore network like the capillary constant and the porosity which are also required for modelling are given in subsection 3.3.3. By means of the capillary constants and the porosities, effective pore radii of the packings were determined (Eq. 4.5) in order to insert the values into the model which are presented in Table C.1 (Appendix C).

From subsection 6.1.2, it is also known how the dissolution of sodium chloride powder affects the load of the solution. Figure 6.16 provides the load versus time plots for the four mixtures with increasing hydrophobic surface in the system. The load is not strongly influenced by the composition of the mixtures. Only in the most hydrophobic sample, the graph for the load of the dissolution is lower. The liquid properties follow the same trend and are presented in Figure C.1 (Appendix C). The load of sodium chloride in water increases to up to $0.42\,g_w/g_s$ within $10\,s$ for the pure sodium chloride sample.

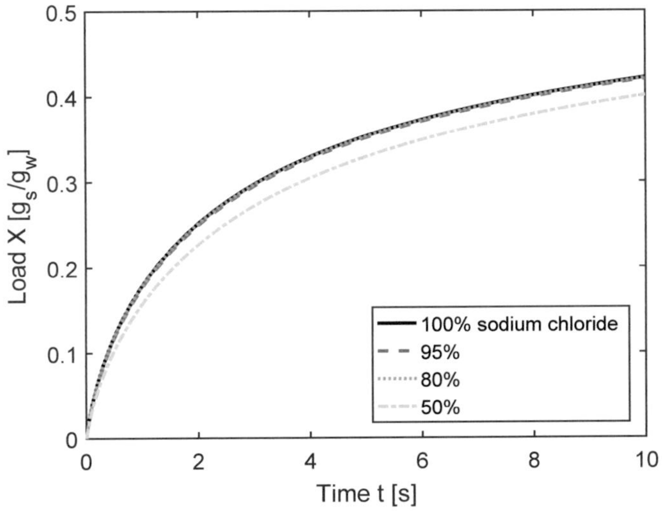

FIGURE 6.16: Load of sodium chloride solution for penetration into systems consisting of sodium chloride and shellac coated glass as a function of time predicted by the model.

As the addition of 5 % and 20 % does not change the load curve significantly, the dissolution is not affected by the small quantities of hydrophobic, inert material. Only in the 50:50-sample, the higher amount of hydrophobic, inert surface reduces slightly the load curve. This effect of an almost constant load curve with decreasing sodium chloride surface in the samples cannot be explained by the dissolution properties, but has to be attributed to the powder properties. A strong increase of effective pore radius accelerates the capillary flow in the model and results in the same load curve for the first three samples. It seems that the reduction of contact area between salt and water in the 95 % and 80 %-sample is balanced by the increased contact area due to a faster penetration flow. In section 6.4, this point is discussed in more detail.

In Figures 6.17 (a) and (b), the squared mass of water which penetrated into the powder mixtures is presented as a function of time determined by experiments and by the model dealing with dissolution. A very clear deviation can be identified by comparing both figures. While the graphs for the four different mixtures differ significantly in case of the experimental plots, the model predicts similar curves for the 100 %, 95 % and 80 %-sample and only a reduced slope for the 50 %-sample. Remembering the findings in subsection 5.2.3, it seems that pore blockage due to an increased contact angle and a resulting reduction of effective porosity leads to the falling slopes of the graphs in Figure 6.17 (a) which is not considered in the model. A detailed comparison of penetration into inert and soluble food systems is presented in subsection 6.2.3. Overall, the model underestimates the liquid penetration into the heterogeneous systems. This can be explained by the neglection of the changing pore network properties due to dissolution which was already discussed in subsection 6.1.2.

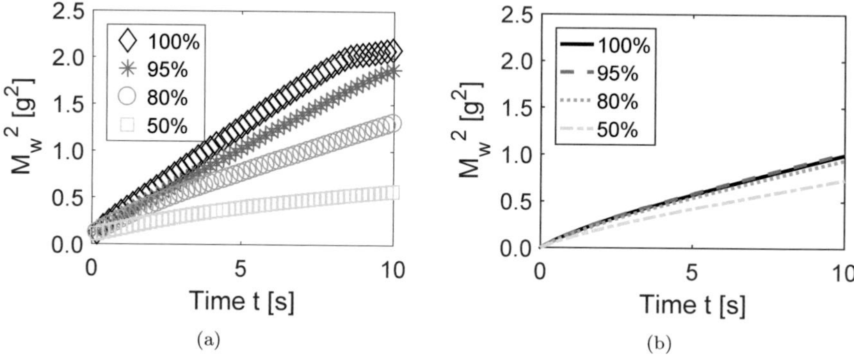

FIGURE 6.17: Squared mass of water during penetration into systems consisting of sodium chloride and shellac coated glass as a function of time determined by experiments (a) and predicted by the model (b).

A precise comparison of experiment and model is given in Figures 6.18 (a) - (d) for each mixture. While a quite good agreement of the graphs can be seen for the two more hydrophobic samples, the deviation inreases in the samples which are more hydrophilic. In the mixtures 80 % and 50 %, the hydrophobic contact angle dominates the penetration flow in the experiment leading to pore blockage. The model generally does not conmsider this aspect but underestimates the penetration. The underestimation due to the neglection of pore network changes and the neglection of pore blockage due to hydrophobicity results in a quite fair agreement of model and experiment. It seems that the two neglected effects balance each other. For the samples 100 % and 95 %, the effect of hydrophobicity caused pore blockage is not as dominant as in the hydrophobic samples. Therefore, it cannot balance the neglection of pore network changes leading to a larger deviation of model and experiment.

Furthermore, the experimental and modelled M^2-t-plots were converted into penetration rates as it was described for sucrose and lactose in subsection 6.1.1 and are presented in Figure 6.19. Inert penetration rates were also calculated by the approach of Benavente et al. (2002) for comparison, as a good agreement of the inert model with experiments was found for pure sodium chloride powder. It can be observed that the two models

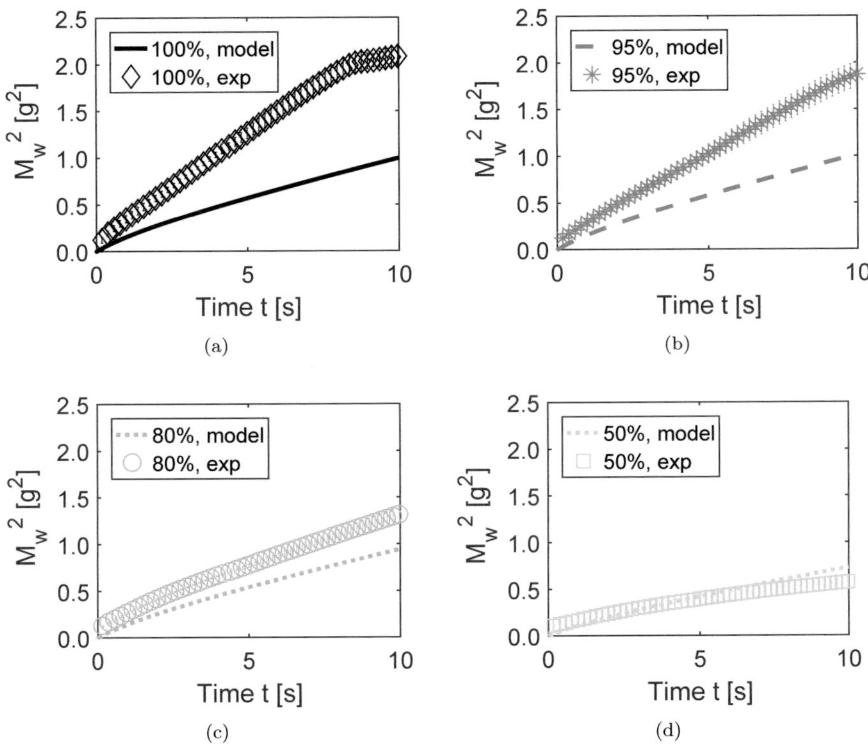

FIGURE 6.18: Squared mass of water as function of time during penetration into systems consisting of sodium chloride and shellac coated glass. Comparison of experiment and model for sample containing 100 % (a), 95 % (b), 80 % (c) and 50 % sodium chloride surface (d).

predict a completely different behaviour for penetration in all four cases. For the two more hydrophilic samples, the inert model is closer to the experiments, while for the two more hydrophobic samples, the new model approximates better the experiments. The reasons for the fair agreement of the new model and hydrophobic samples was already explained above in the discussion for Figures 6.18 (a)-(d). In case of the agreement of the inert model with the hydrophilic samples, the explanation was presented in subsection 6.1.2 for pure sodium chloride powder. As long as the hydrophobicity is not dominant, the two effects of dissolution which are the change of liquid properties and the change of pore network, seems to be balanced. Thus, the inert model neglecting dissolution effects show a fair agreement with experimental data.

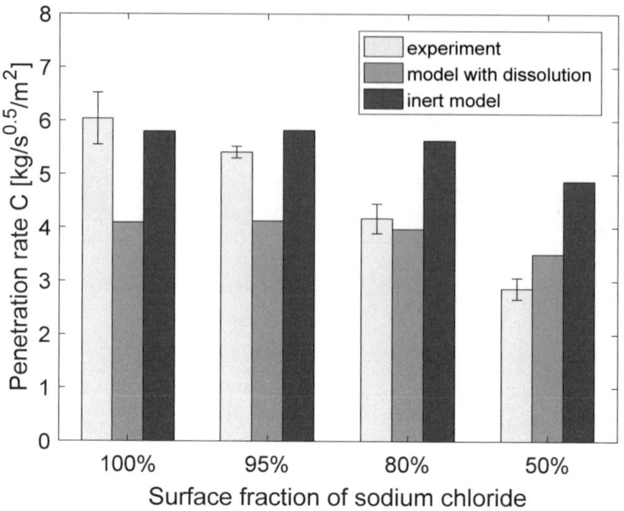

FIGURE 6.19: Penetration rates into four mixtures of sodium chloride and shellac coated glass determined by experiments, with the developed model for dissolution and with an inert model.

6.2.3 Comparison of food with inert systems

In this subsection, the results and conclusions from investigating the liquid penetration into heterogeneous, inert systems (section 5.2) are compared to the findings from this chapter. The replacement of the inert, hydrophilic component by a hydrophilic food component leads to the increase of complexity since dissolution and its consequences have to be considered. Generally, two heterogeneous food cases with different dissolution behaviour were investigated. The system with sucrose is mainly influenced by the viscosity development during liquid penetration, while there is not only one dominant effect in case of the salt system. The alteration of liquid properties and the change of pore network influence the penetration behaviour into sodium chloride in a balanced way. Hence, the hydrophobic contact angle is dominant for the wetting performance into the heterogeneous salt system which is also the case for the inert system.

Figure 6.20 provides a direct comparison of the penetration rates into the heterogeneous sucrose system and the inert system. The rates are plotted as a function of the cosine of the Cassie-Baxter contact angle as an expression of the hydrophobicity of the mixture. A completely different behaviour can be observed for the sucrose and the inert case. While without dissolution the penetration rate continously decreases with increasing hydrophobicity (decreasing cosine of Cassie-Baxter contact angle), the rate decreases only for the most hydrophobic mixture in case of the soluble sucrose system. For low and

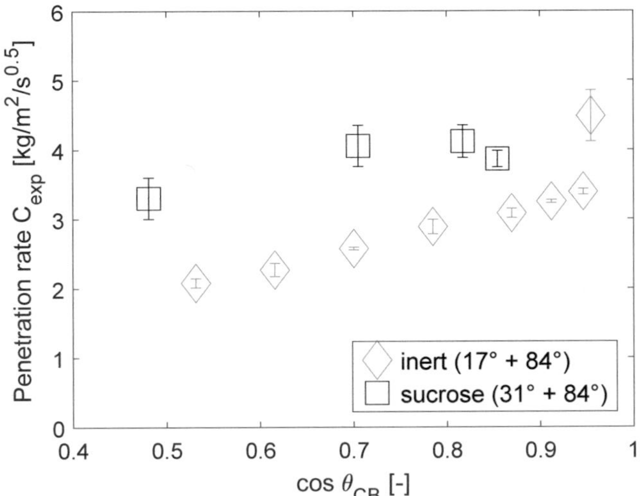

FIGURE 6.20: Comparison of penetration rates of heterogeneous, inert system ($17°$ + $84°$) and heterogeneous, food system ($31°$ + $84°$) plotted as a function of the cosine of the Cassie-Baxter contact angle.

intermediate quantities of hydrophobic component in the sucrose system, the penetration rate is not influenced. As it was already explained in subsection 6.2.1, two effects are balanced in this concentration range. The addition of hydrophobic surface increases the contact angle, but reduces also the contact area between sucrose and water, thus, the dissolution and the increase of viscosity. The overall higher penetration rates for the sucrose system are related to the powder properties. The liquid flow into the inert system of coarse glass beads is driven by a capillary constant of $1.8\,\mathrm{mm}^5$ and the sucrose-mixtures exhibit capillary constants from $3.7\,\mathrm{mm}^5$ to $3.9\,\mathrm{mm}^5$. The glass beads with a very high sphericity of 0.92 can build a closer packing than the mixtures containing sucrose with a sphericity of 0.84. Larger pores in the looser powder bed favour a fast liquid penetration.

By contrasting penetration into the inert system with penetration into the sodium chloride mixtures in Figure 6.21, a similar trend for the rates can be observed. With increasing hydrophobicity (decreasing cosine of Cassie-Baxter contact angle), the penetration rates decrease continously.

Due to the compensation of the main two dissolution effects, the trend of the penetration behaviour into the salt mixture is comparable to penetration into the inert system. As it was already seen for the sucrose mixtures, the penetration rates are generally higher than the ones of the inert system which is attributed to the powder properties. The capillary constants of the salt mixtures are with values from $3.6\,\mathrm{mm}^5$ to $3.8\,\mathrm{mm}^5$ in the

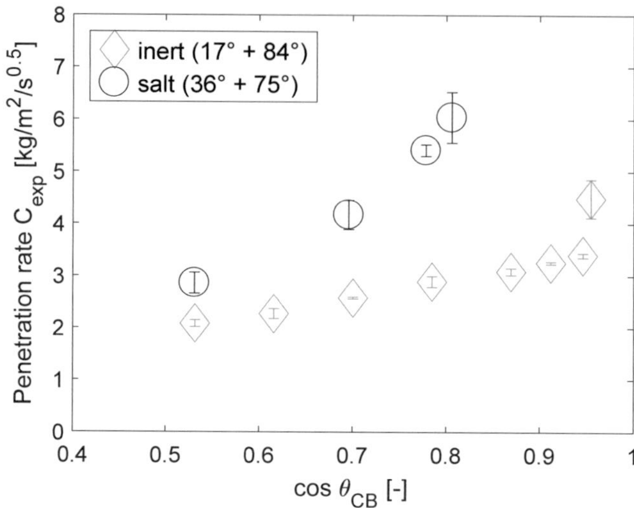

FIGURE 6.21: Comparison of penetration rates of heterogeneous, inert system (17° + 84°) and heterogeneous, food system (36° + 75°) plotted as a function of the cosine of the Cassie-Baxter contact angle.

same size range as the sucrose mixtures and about twice the value of the one of glass beads. The lower sphericity of sodium chloride of 0.80 leads to the formation of a loose pore network for penetration.

In subsection 6.2.2, it could be seen that the application of the inert model resulted in penetration rates very close to the experiments in case of the more hydrophilic samples. Therefore, in Figure 6.22 (a), the experimental rates are again compared to the rates modelled with the inert approach. As it was already seen in Figure 6.19, the agreement of model and experiment is only given for the more hydrophilic sample. With increasing hydrophobicity, the deviation between model and experiment increases. Due to the similarity of salt mixtures and inert mixtures, the porosity adaption which was applied in case of the inert powder, is transferred to the salt systems. The measured porosities of the four salt systems are 0.50, 0.49, 0.46 and 0.40 with increasing quantity of glass beads in the system. An adaption to porosity values of 0.50, 0.46, 0.34, 0.24 results in the modelled graphs in Figure 6.22 (b). It seems that also for the salt mixtures, an increasing amount of hydrophobic surface in the system results in the reduction of accessible pore volume for water. A partial pore blockage is the consequence if pores are mainly surrounded by hydrophobic beads. In case of the most hydrophobic mixture, the pore volume is reduced by 40 % as it was also observed for the most hydrophobic mixture of the inert system.

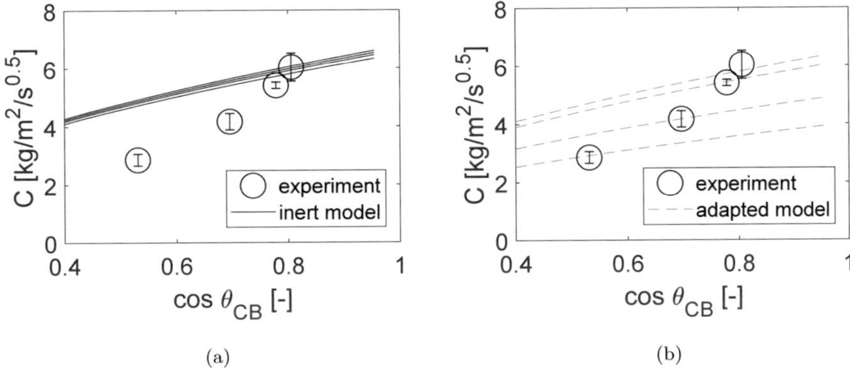

FIGURE 6.22: Penetration rates for heterogeneous system of sodium chloride and shellac coated glass beads; comparison of experiment and inert model (a) and comparison of experiment and inert model with adapted porosity (b).

6.3 Combined factors in real food system

Besides the investigation of penetration into heterogeneous and soluble model food systems, the wetting performance of two real food powders was studied with experiments and compared to modelled results. Therefore, the penetration behaviour into two milk powders, a standard whole milk powder (WMP) and a whole milk powder with free surface fat (WMP ff) which are described in more detail in section 3.1 were recorded. First of all, it could be seen that no penetration was found in case of the WMP ff. Even if the bulk composition of WMP and WMP ff was identical the coverage of the particle surface with fat by 80 % resulted in dynamic contact angles on a rough surface of 114.4° to 100.7° leading to no capillary rise of water into the powder network. This finding emphasizes the importance of the location and the distribution of hydrophobic components in a food particle. Hereinafter, merely the results for WMP are presented and discussed while WMP ff is not further considered.

In Figure 6.23, the load of solution during the penentration of liquid into WMP and lactose is presented as a function of time. The graph was calculated by the model.

It can be seen that the solution load for WMP increases up to a value of $0.09\,\mathrm{g_s/g_w}$ within $6\,\mathrm{s}$ while for lactose the value rises slightly higher to $0.11\,\mathrm{g_s/g_w}$. For modelling the penetration into WMP, the dissolution parameters were taken from lactose since it is assumed that lactose is the main soluble component in milk powder. Thus, the load dependency of liquid properties during the dissolution of lactose was taken from Figure 6.1. Furthermore, the lactose values for the hydrodynamic radius and the saturation

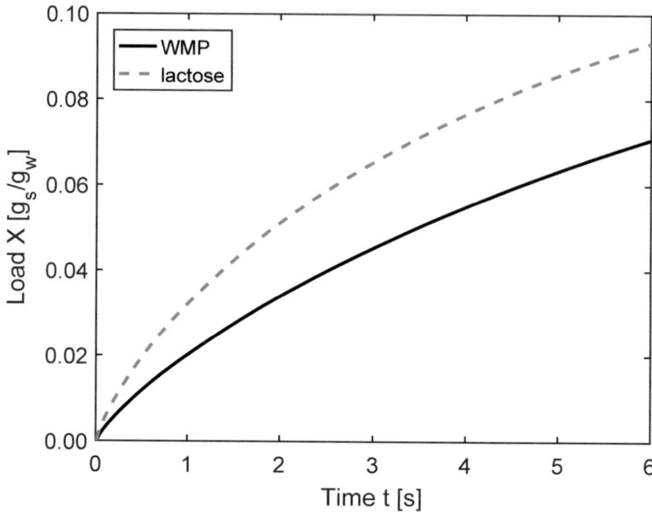

FIGURE 6.23: Load of lactose solution for penetration into WMP and into lactose powder as a function of time predicted by the model.

density were inserted in the model for WMP. Other parameters for modelling were experimentally determined for WMP. The dynamic contact angle for water on WMP was taken from Figure 3.18 whereby the contact angle of 82.3° at the most dynamic point (contact line speed of $6.0\,\mathrm{mm\,s^{-1}}$) was inserted in the model. All powder properties which are required for the model are summarized in subsubsection 3.3.2.2 such as the porosity, the capillary constant, the effective pore radius and the mean particle diameter. As the dissolution properties of WMP and lactose are comparable due to the same dissolving component, the differences of the load can be attributed to the powder properties. This aspect is explained more detailed when discussing the squared mass of water as a function of time.

The comparison of experimental and modelled data for the squared mass of water as a function of time are shown in Figure 6.24. It can be seen that the model starts to overestimate the experimental data after about $1.5\,\mathrm{s}$.

This deviation can be explained by two aspects. First of all, as it was already discussed earlier, the change of the pore network due to dissolution is not considered in the model. Therefore, blockage of pores caused by partial collapse which happens in the experiment does not influence the modelled data. Secondly, there are more soluble components besides lactose in WMP such as proteins. The dissolution of those components also affects the liquid properties and is not considered in the model. Anema et al. (2004) observed that the viscosity increase during dissolution of milk powders depends on the formation

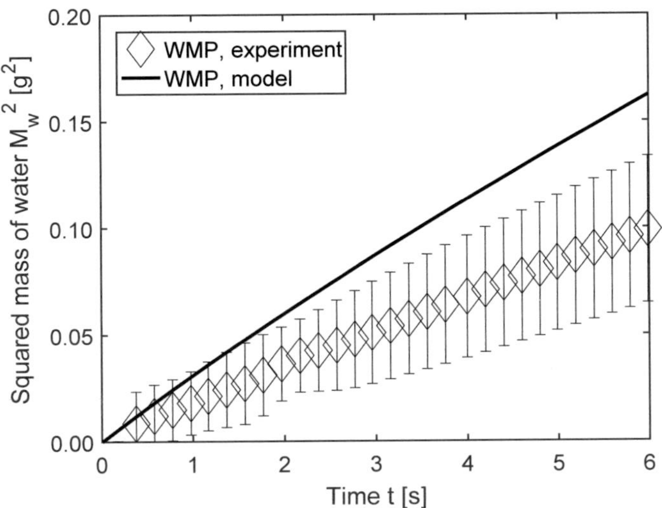

FIGURE 6.24: Squared mass of water as function of time during penetration into milk powder (WMP): experimental data in comparison to data predicted by the model.

of casein micelles. Furthermore, the presence of dissolved protein reduces the surface tension of the solution affecting also the wettability (Yang et al., 2008). Nevertheless, the predicted liquid penetration into a real food powder which exhibits heterogeneity and solubility is already close to the experimental one. This statement can be emphasized when determining penetration rates of the experimental and the modelled data. Penetration rates of $1.59\,\mathrm{kg\,m^{-2}\,s^{-0.5}}$ (experiment) and of $2.04\,\mathrm{kg\,m^{-2}\,s^{-0.5}}$ (model) were calculated which are quite close to each other. For comparison, the inert approach predicts a value of $2.44\,\mathrm{kg\,m^{-2}\,s^{-0.5}}$.

Figure 6.25 provides a comparison of the modelled squared mass of water as a function of time graphs for WMP and lactose. Since for both systems the dissolution is described by the dissolution parameter of lactose, the behaviour of the liquid properties viscosity, density and surface tension during penetration into WMP is similar to the one of lactose which is shown in Figure 6.3. Hence, the deviation between the graphs in Figure 6.25 can be attributed to the contact angle and to the properties of the pore network.

The dynamic contact angle of WMP with 82.3° is clearly larger than the one of lactose with 24.4° which favours the penentration into lactose powder. A further improvement of penetration into lactose compared to WMP is expressed by the capillary constants. The value of $5.9\,\mathrm{mm^5}$ for lactose is strongly larger than the value of $3.8\,\mathrm{mm^5}$ for WMP. Consequently, the effective pore radius which is inserted into the model and which influences the capillary flow is also larger for lactose (5.7 μm) compared to WMP (3.0 μm).

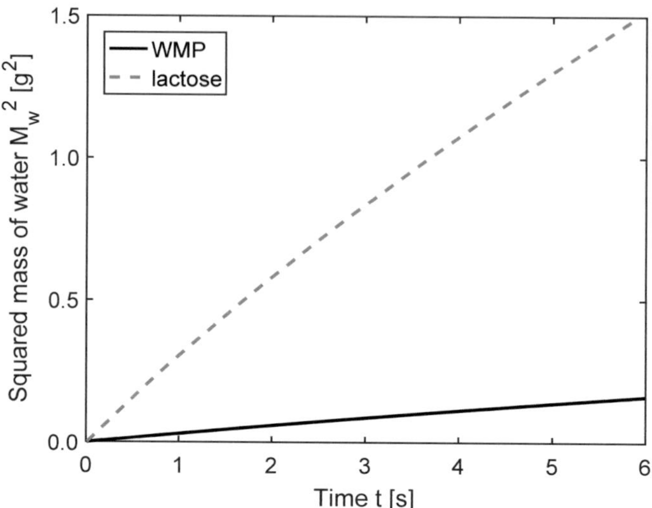

FIGURE 6.25: Squared mass of water as function of time during penetration into milk powder (WMP) and lactose predicted by the model.

If the pore properties or the contact angle is the main decelerating factor in the case of WMP is not clear at this point. But in order to answer this question and to gain understanding about the weighing of the different influencing factors a detailed parameter study follows in section 6.4.

6.4 Model for prediction of capillary wetting

In this section, the new model for heterogeneous, soluble food systems is used to perform a detailed parameter study in order to find out how to improve the wettability by different influencing factors. Secondly, proposals are made to enhance the model towards a better agreement with experimental data.

6.4.1 Numerical study of the sensitivity of wetting factors

Since the new model for wetting with dissolution provided only a reliable prediction for penentration into the systems with sucrose, in this subsection, a parameter study will be performed for a heterogeneous, soluble system where viscosity increases is the dominant influencing factor. The parameter study is separated into three study cases in order to find out how the wetting is affected. Furthermore, it is aimed to give a recommendation how the penetration into a heterogeneous, soluble food powder can be enhanced.

For performing the parameter study, all parameters from pure sucrose powder were taken as standard values. The dissolution is characterized by the hydrodyanamic radius, the saturation density and the mean particle diameter of sucrose. Furthermore, the liquid properties depending on the load, the contact angle and the pore network properties of sucrose were used. The proportion of contact area between liquid and soluble solid was set as 1 for the standard case. For each case, only one parameter was varied while the other parameters were kept constant. The results are expressed by the load and the squared mass of water as functions of time.

6.4.1.1 Influence of contact angle

In a first study, the influence of the contact angle on the dissolution and the penetration was investigated. Therefore, the contact angle was varied from $0°$ for a perfectly wetting system to $90°$ for a non-wetting system in interval steps of $10°$. The modelled load versus time and the modelled squared mass of water versus time are presented in Figure 6.26 (a) and (b), respectively, for the ten different contact angles.

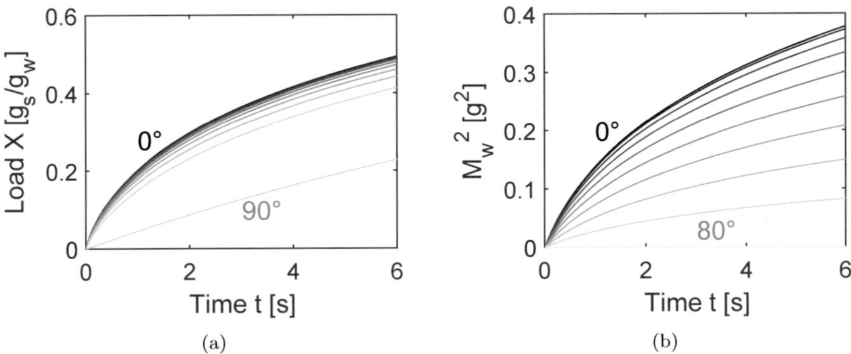

FIGURE 6.26: Load (a) and squared mass of water (b) as functions of time predicted by the model for contact angles ranging from $0°$ to $90°$.

The load does not strongly depend on the contact angle as long as the height depending contact area between solid and liquid for dissolution is kept constant. Only a contact angle of $90°$ which inhibits capillary penetration changes the load curve significantly. Taking Figure 6.26 (b) into account, it can be seen that for the $90°$-case, no penetration can be observed (light grey line on x-axis). Therefore, no new contact area is created by an upwards flow and only a small constant contact area is available for dissolution. Comparing the curves in the M^2-t-plot, it is obvious that the contact angle has a strong influence on the penetration flow. With an increasing contact angle the liquid penetration is reduced. The decreasing trend is not linear with increase of contact angle since the capillary flow depends on the cosine of the contact angle. Therefore, the relative reduction of liquid penetration increases with increasing contact angles.

Even if the penetration is reduced for higher contact angles, the liquid within the pore network has relatively the same contact area with the soluble component and, thus, the load does not change intensively. The small variations are explainable by a higher contact angle decreasing the penetration velocity. A lower penetration velocity reduces the Reynolds number and the Sherwood number which again decreases the mass transfer coefficient. It can be concluded that the liquid penetration is significantly affected by an increasing contact angle when the height depending contact area stays constant. On the other hand, the relative dissolution is not strongly influenced by an increasing contact angle except in the case of a contact angle of 90°.

6.4.1.2 Influence of dissolution area

In a next stept, the height depending contact area between liquid and soluble material for dissolution is varied. In real food system, this can be the case if non soluble material such as fat is present in the powder. Generally, with the presence of a fatty, non soluble component in the powder also the overall contatc angle will increase. But in order to understand the effect of each parameter, the influence of contact area was studied separately.

Figure 6.27 provides the load (a) and the squared mass of water (b), both as a function of time predicted by the model for contact areas between liquid and soluble solid ranging from 100 % to 0 % with a step width of 10 %. It is immediately obvious, that two different trends are obtained for dissolution and penetration by a changing contact area.

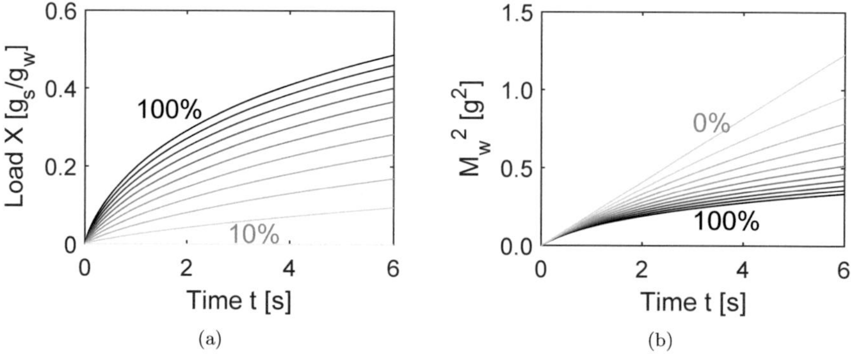

FIGURE 6.27: Load (a) and squared mass of water (b) as functions of time predicted by the model for contact areas ranging from 100 % to 0 %.

The dissolution and, thus, the load of sucrose decreases significantly with decreasing height depending contact area. For a contact area of 0, no sucrose can dissolve in the liquid (curve on x-axis). On the other hand, the penetration flow increases with decreasing contact area which can be attributed to less viscosity increase with less dissolution. By reducing the contact area from 100 % to 0 %, a relative increase of squared mass of water of 270 % is obtained after 6 s. Thus, it can be concluded that the contact area for dissolution has a strong influence on the load and on the penetration curve. The dissolution is accelerated with a higher contact area which reduces the penetration flow due to viscosity increase.

The influence of contact angle and of contact area for dissolution are usually connected with each other, since a reason for an increased contact angle is often the presence of fatty, non soluble componts which reduces the contact area. Hence, the effect of the two parameters is compared. The opposite trend can be observed when comparing Figures 6.26 (b) and 6.27 (b). An increased contact angle decreases the penetration flow strongly, while a simultaneous reduction of dissolution area increases the penetration. This emphasizes the trend in Figure 6.13 where an increasing contact angle did not show a strong influence on penetration.

6.4.1.3 Influence of pore network properties

Besides contact angle and dissolution area, the pore network properties influence the penetration into a powder. Since the penetration in the model is driven by capillary forces, the capillary radius is also the main pore network parameter influencing the flow. The pore network properties were characterized by the determination of the capillary constant which is converted into an effective pore radius by means of the porosity. In this parameter study, the effective pore radius was varied from $1\,\mu m$ to $8\,\mu m$ in steps of $1\,\mu m$ which is in accordance to the range of calculated values in this thesis. In Figure 6.28, the load versus time (a) and the squared mass of water versus time (b) for the seven different effective pore radii are plotted.

The load curves are slightly affected by the increasing pore radius when keeping all other parameters constant. After 6 s, the load increases by 22 % when increasing the pore radius from $1\,\mu m$ to $8\,\mu m$. On the other hand, the pore radius shows a strong impact on the penetration rate. With an increases of pore radius from $1\,\mu m$ to $8\,\mu m$, the squared mass of water increases by 533 % after 6 s. The trends achieved by the variation of pore radius are similar to the ones achieved by contact angle variation. Hence, the low influence on the load can be explained in the same way. The liquid within the pore network with a radius of $1\,\mu m$ rises slower but has in relation to its mass the same

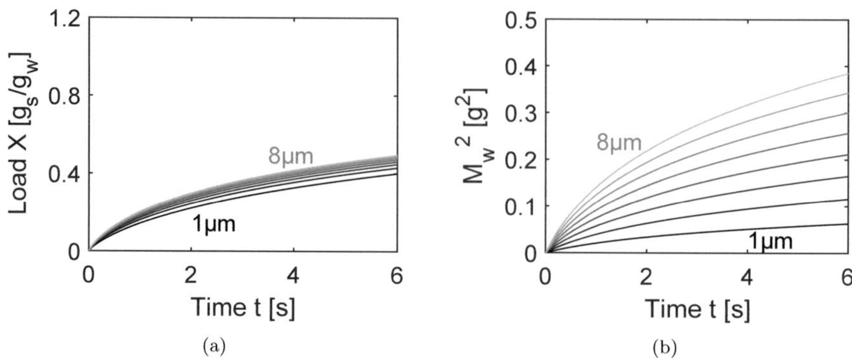

FIGURE 6.28: Load (a) and squared mass of water (b) as functions of time predicted by the model for contact areas ranging from 1 µm to 8 µm.

contact area with the soluble component than the faster rising liquid in the 8 µm-system. The marginal variations can be attributed to the penetration velocity and its influence on the mass transfer coefficient.

As in reality, a changing size of pores in a powder system is always linked to a changing particle size, both parameters were also combined in a parameter study. For each pore size, a related particle size was assigned. The particle sizes ranges from 50 µm to 330 µm with a step width of 40 µm. The smallest particle size belongs to the smallest pore size. In Figure 6.29, the load and the squared mass of water are plotted as functions of time for the seven different cases .

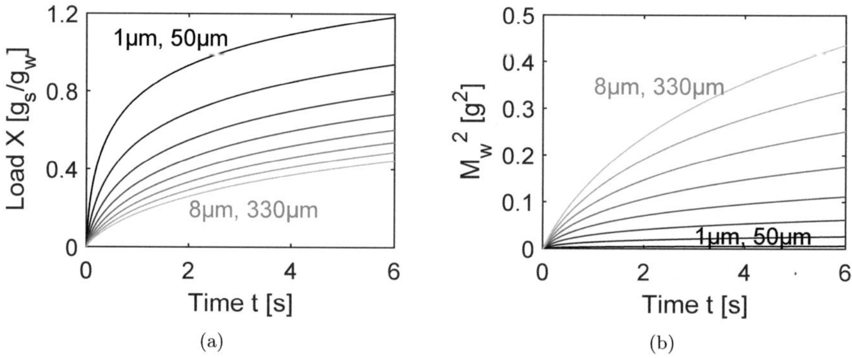

FIGURE 6.29: Load (a) and squared mass of water (b) as functions of time predicted by the model for contact areas ranging from 1 µm to 8 µm and particle diameters ranging from 50 µm to 330 µm.

A significant change of the trend for the load curves compared to Figure 6.28 (a) is obtained when the alterating particle size is considered. The load after 6 s decreases by 64 % with the variation of pore and particle size from minimum values to maximum values. The smaller the particle size, the higher is the contact area for dissolution due to an enlargement of particle surface. The influence of particle size is significantly stronger than the effect of pore size on the load curve. The trend for the penetration curve remains the same with a broadening of the array of curves. The penetration flow for the smallest pore size is further reduced due to a higher viscosity increase caused by increased dissolution. For the largest pore size, the dissolution rate is slightly decreased by an increased particle size which results in a small increase of the penetration curve.

In the end, the influence of pore network properties on dissolution and penetration is summarized. While the pore size showed a strong and direct effect on the penetration behaviour, the load is almost independent of this parameter. On the other hand, the main influence of the particle size is on the dissolution rate and, thus, on the load of the solution. The penetration is only indirectly influenced as dissolution influences the liquid properties and, therefore, the penetration behaviour.

6.4.2 Approaches for improvement of the model

As a last step of this chapter, some aspects are suggested, decribed and discussed which can probably enhance the performance of the new model towards a better prediction of the reality. Therefore, three deficiencies within the model were selected and tried to improve. Since the best agreement between experiment and reality was achieved in case of the sucrose systems, the sodium chloride powder and the milk powder were chosen for improvement studies. In case of sodium chloride powder, the effect of a dynamic contact angle and of the changing pore network due to dissolution were tested. The modelling of the wettability of milk powder was modified by adaption of the liquid properties during dissolution.

In Figure 6.30, the effect of contact angle within the observed range of dynamic contact angles for sodium chloride is presented. The squared mass of water as function of time was calculated by the new model with the maximum contact angle of 36.3° measured for sodium chloride. Furthermore, the graph was plotted for the minimum contact angle of 19.2° in order to see what is the impact on the penetration behaviour. Comparing the two calculated graphs with the experimental results, it is obvious that the lower contact angle shifts the graph only slightly towards the experimental plot and, thus, does not provide a satisfactory improvement. It can be concluded that an implementation of the dynamic contact angle into the model would not result in a sufficiently better prediction.

Besides the effect of the dynamic contact angle, the influence of the changing pore network was also studied as a possible opportunity for improvement. Therefore, two

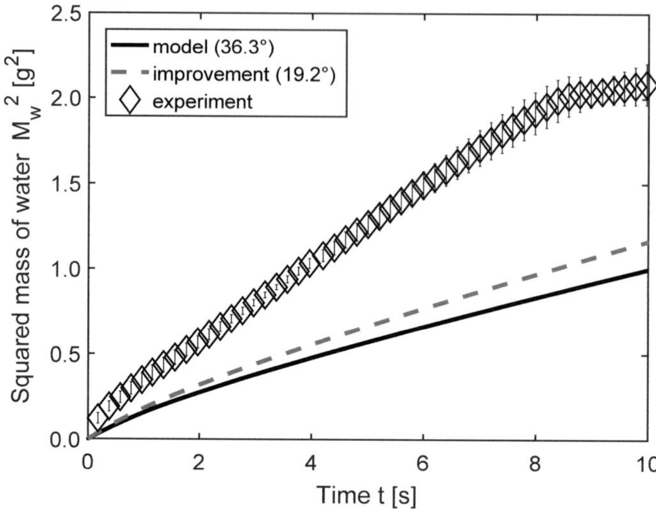

FIGURE 6.30: Squared mass of water as function of time during penetration into sodium chloride predicted by the model with a contact angle of 36.3° (standard), predicted by an improved model with a contact angle of 19.2° and experimentally determined.

different options, improvement 1 and 2, were tested for determination of the change of pore network properties and particle size on the penetration behaviour. In the course of improvement 1, the dissolution of a single sodium chloride particle was recorded for 330 s as it is shown in Figure 6.31. The images were evaluated to experimentally determine the decrease of the projection surface of the particle which was used to determine the volume by assuming spherical beads. Subsequently, the volume reduction was used to calculate the change of three relevant parameters in the model: the decrease of an equivalent particle diameter, the increase of porosity which is equal to the pore volume and the increase of pore radius. For modification of the parameter, the values after 10 s were taken. During this time period the particle diameter decreased from 288 µm to 281 µm, the porosity increased from 0.50 to 0.53 and the pore radius is increased from 4.7 µm to 4.8 µm. The model was run again with these values. More details about the determination of the different parameter are provided in the Appendix D.

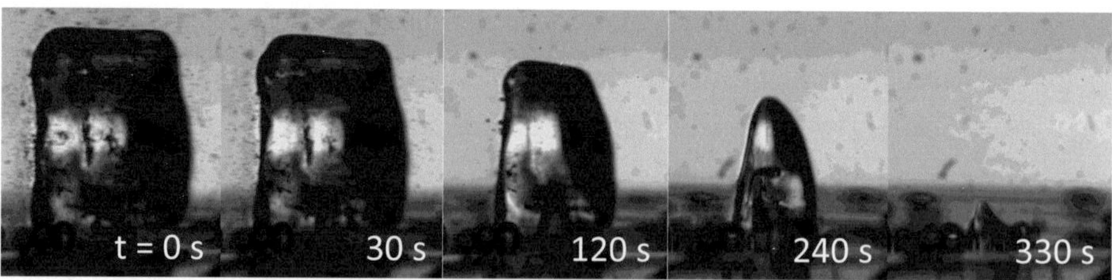

FIGURE 6.31: Images of a dissolving sodium chloride particle during 330 s.

In a second improvement approach, the particle size, the porosity and the pore radius were modified by means of the dissolution data obtained from the model (improvement 2). Therefore, the mass of dissolved sodium chloride after 10 s was taken from the results and used for the calculation of the reduction of particle volume. Consequently, the volume was converted into the particle diameter. Additionally, the ratio of reduced particle volume and diameter were set to be equal to the increase of porosity and pore radius, respectively. The new parameters obtained by the approach are 268 µm for the particle diameter, 0.60 for the porosity and 5.0 µm for the pore radius. Afterwards, the model was run again with the newly determined parameters.

In Figure 6.32, the results of the two improvement approaches are shown in comparison to the graph obtained by the existing model and the experimental values. It can be seen that applying the improvement approaches, the graphs shift towards the experimental data, while improvement 2 achieved a better result. This can be attributed to the larger change of particle size, pore radius and porosity in the second approach. Since the dissolution experiment for approach 1 was performed under static conditions in terms of water flow, the dissolution kinetics might not be applicable for the dynamic conditions during liquid penetration. But since the dissolution data which was taken from the existing model for approach 2 was modelled without considering the effect of changing particle size and pore network, this approach contains an error. Thus, it is recommended to redo the dissolution test under dynamic conditions in order to obtain more reliable experimental results for the dissolution kinetics.

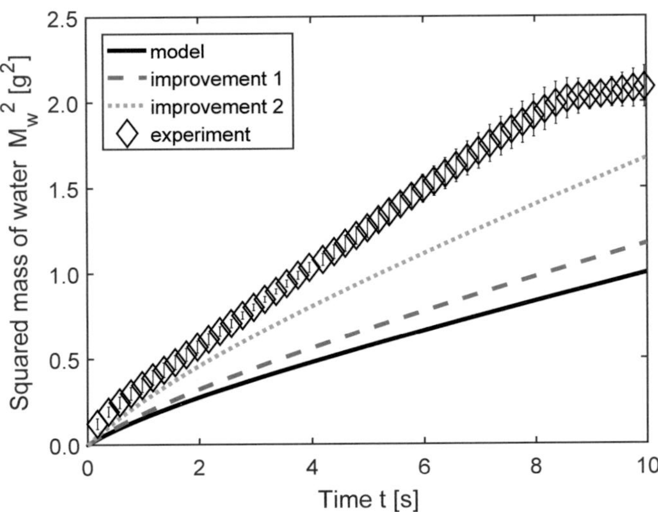

FIGURE 6.32: Squared mass of water as function of time during penetration into sodium chloride predicted by the model, by two improved calculations (improvement 1 and 2) and experimentally determined.

As a last action, an improvement approach for predicting the penetration into milk powder is presented. Since milk powder is very heterogeneous in its composition, only one component, lactose, was considered for dissolution in the existing model. However, there are more components besides lactose which dissolve and have an impact during wetting of milk powder such as proteins. In Figure 6.33, the viscosity as a function of load is presented for lactose and whey protein isolate as an example. Whey protein is one of many proteins being present in milk. During its dissolution, it creates a higher viscosity development in a solution at the same concentration level compared to lactose. Thus, considering additionally the dissolution of proteins would most probably lead to an enhanced viscosity increase and, therefore, to a decreased penetration. The different dissolution kinetics and different effects on the liquid properties of each dissolving component make it highly complex to introduce it into the model. Hence, it is recommended to figure out which is the dissolving component affecting the liquid properties the most and introduce it into the model.

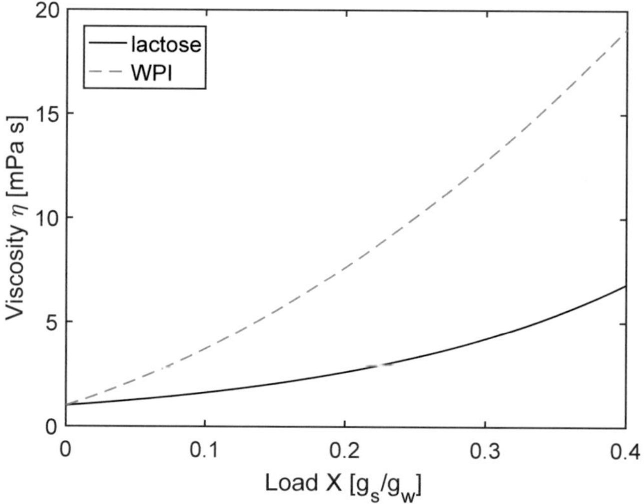

FIGURE 6.33: Viscosity as a function of load for two solutions of lactose and whey protein isolate (WPI) (Patocka et al., 2006, Walstra, 1999).

6.4.3 Application of model from an industry perspective

Over the last century, the consumer preferences towards food products have changed. Nowadays, the consumer is more interested about the origin, the production process and the composition of the products. Unfamiliar ingredients and some processing methods are considered as not healthy and not natural by the consumer. This consumer perception is expressed as clean label trend (Asioli et al., 2017). Other growing consumer trends are vegetarianism and veganism. Both are related to a plant based nutrition which is considered as healthier and more sustainable compared to a diet with meat (North, 2017, van Loo et al., 2017). All these trends have to be understood and implemented in products. Therefore, the food industry is challenged by renovating their existing product portfolio and developing new products.

The consequence for the food powder sector is the need of launching new powders such as fruit powders or plant-based milk powders fulfilling the consumer expectations (Jeske et al., 2018, Zotarelli et al., 2017). Furthermore, existing recipes have to be revised in order to replace ingredients which are not longer accepted by the consumer (Asioli et al., 2017). However, consumers are used to powders with high level properties such as fast reconstitution. Thus, the novel powder generation has to show a comparabel behaviour.

The model for prediction of capillary wetting of heterogeneous powder developed in this thesis can support the study of the first step of reconstitution of novel food powders and renovated recipes containing new components. Especially if the powders are rich in carbohydrates and viscosity build-up might be an issue a parameter study by means of the model is useful for defining a suitable surface composition of the particles. Furthermore, the model can help to identify the proper particle size which supports the capillary wetting.

7

Conclusion

In this chapter, the main results of the thesis are summarized and the conclusions are emphasized. In the first section, the key findings obtained in inert single gaps and powder systems are explained. The second section focuses on the results which were achieved in homogeneous and heterogeneous food systems.

7.1   Inert systems

In the first part of this thesis, the capillary penetration into heterogeneous, inert systems of glass beads was investigated. It was found that with an increasing hydrophobic content the wetting performance decreases. At a critical concentration of 1 % hydrophobic surface the penetration rate exhibits a sharp reduction. For a small and a coarse fraction the same trend could be observed with an overall faster penetration into the coarse powder.

When results obtained in single gaps are compared to the results of powder systems, the influence of the pore network on the wetting behaviour gets obvious. While the contact angle of the hydrophobic component showed a significant effect on the capillary rise in a single gap, the penetration into a heterogeneous pore network is not affected by a higher hydrophobic contact angle. The reason for this behaviour could be clarified by applying a model for the penetration into inert systems and variation of the porosity of the pore network. The presence of an accumulated hydrophobic surface can cause the blockage of pores leading to a reduced effective porosity which are the open pores for penetration. Since water does not enter those highly hydrophobic pores, a higher contact angle does not have an effect. A decrease of effective porosity of 40 % was found by addition of 50 % of hydrophobic surface in the fine and the coarse powder.

7.2   Soluble systems

In the second part of this thesis, the liquid penetration into soluble, homogeneous and soluble, heterogeneous systems was studied. The knowledge gained with the inert systems was used for understanding heterogeneity effects in the soluble systems. A new model was developed for the prediction of liquid penetration into soluble and heterogeneous powders dealing with dissolution during wetting. The model is based on the Washburn equation for capillary penetration and on the mass transfer equation for dissolution. Heterogeneity in terms of hydrophilic and hydrophobic components is added by calculation of mixed contact angles. The dissolution during water uptake influences the liquid properties but does not affect the pore network properties.

Performing wetting experiments and applying the model different findings were obtained for sucrose, lactose and sodium chloride. In case of sucrose powder which is a fast dissolving material, the viscosity development during wetting was identified as main influencing factor. The modelled penetration rate was in fair agreement with the experimental results. Thus, the model was verified for predicting liquid penetration into sucrose. For lactose as slower dissolving powder, a dominant effect other than viscosity development is driving the penetration. It seems that the change of pore network with a resulting partial bed collapse influenced the water penetraton into the lactose powder. Since changing pore network properties are not considered in the model, the agreement of experiment and model is not as significant as for sucrose. Using four size fractions of sodium chloride powder allowed the investigation of the effect of particle size. Both, experiments and model confirmed an optimum of particle size for liquid uptake. The penetration rates are reduced in finer and coarser powders compared to the optimum. Only a weak agreement of experiment and model could be achieved due to neglection of changing pore network properties due to dissolution. Since the effect of changing liquid properties and changing pore network properties seem to balance each other, an inert model taken from literature showed a better agreement with experimental results. In a parameter study, it could be confirmed that the alteration of both, pore radius and particle size, have an intense effect on the dissolution and liquid penetration.

Since food powders often do not consist of a single component, the wettability of mixtures containing a stochastic distribution of hydrophilic, soluble particles and hydrophobic, inert particles was investigated and compared with predicted wetting performance by the developed model. In systems made of sucrose and hydropbic glass beads, two main effects, the viscosity development and the hydrophobic contact angle, drove the liquid penetration. With increasing hydrophobic content, the contact area between liquid and the soluble component is decreased leading to less dissolution and, thus, to less viscosity increase. This aspect improves the wetting performance, while an increased hydrophobic content also increases the overall contact angle which again reduces the liquid penetration. There is an optimum for the wetting performance at low concentrations of hydrophobic component. The effect of competing factors on dissolution and penetration is emphasized in a parameter study. Penetration rates predicted by the new model were in good agreement with experimental rates. Hence, the model was also verified for predicting liquid penetration into mixtures of stochatic distributions of sucrose and hydrophobic particles.

In a second heterogeneous system containing sodium chloride powder (hydrophilic, soluble) and shellac coated glass beads (hydrophobic, inert), the hydrophobic contact angle was detected to be the dominant influencing factor. Since the increase of viscosity during dissolution is reduced compared to sucrose, there is no competing effect due to dissolution. Thus, a similar trend compared to the inert systems could be observed.

As a final step, the developed model was applied for milk powder which is a real food powder consisting of different components and, therefore, exhibits heterogeneity in terms of solubility and contact angle. It was found that the model overestimated the penetration into milk powder when comparing with experiments. Since for dissolution only the component lactose was implemented, the effect of protein dissolution, for instance, is neglected in the model. Due to high complexity in composition the prediction of wettability for food powders such as milk powder is challenging. Anyway, the modelled results provide already a fair approximation of the wetting behaviour. An improvement of prediction could be achieved by inserting the dissolution kinetics and its consequences for liquid properties of all dissolving components.

This thesis provides an improved understanding of the liquid penetration into heterogeneous and soluble food powders during reconstitution and offers a new model for predicting the water uptake into such systems.

In course of this thesis some deficiencies of the model could be noticed. For improving the performance of the new model, a detailed consideration of each influencing factor has to be added which were neglected so far. It seems that the incorporation of the effect of dissolution on the pore network and the particle size is essential for systems which are not dominated by viscosity development. For complex systems with more than one dissolving component, the dissolution kinetics of all main dissolving components have to be considered. So far all powders were treated as monomodal particles. An implementation of a particle size distribution would change the pore network properties towards reality. Furthermore, the developed model which is a tool for predicting the capillary penetration into powder systems should be combined with other models predicting the sinking, the dispersion and the dissolution to create a method to predict the whole reconstitution process.

Appendix A

Sodium chloride

Figure A.1 shows the surface tension as a function of time predicted by the model for four different size fractions of sodium chloride. It can be seen that the surface tension increases with increasing time as it was also observed for sucrose. The increase of particle size from s_1 to s_4 results in a decrease of surface tension which was determined for the other liquid properties.

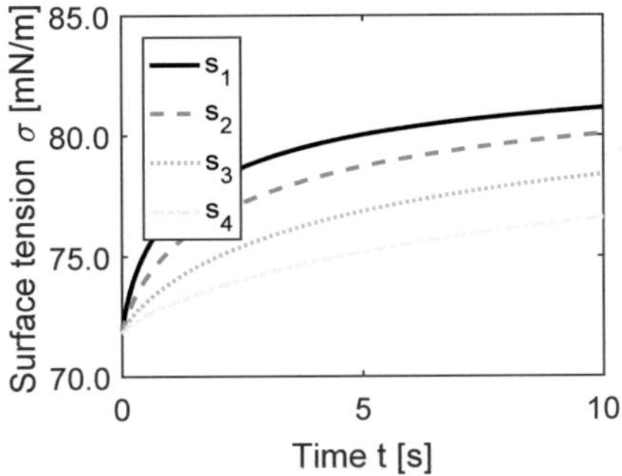

FIGURE A.1: Surface tension as a function of time for four sodium chloride fractions predicted by the model.

Secondly, the Washburn plots (M^2-t-plots) are presented for the four different fractions in Figure A.2 (a) - (d). In each diagram, the experimental results are compared to the modelled data. For all four cases, the model underestimated the experimental results.

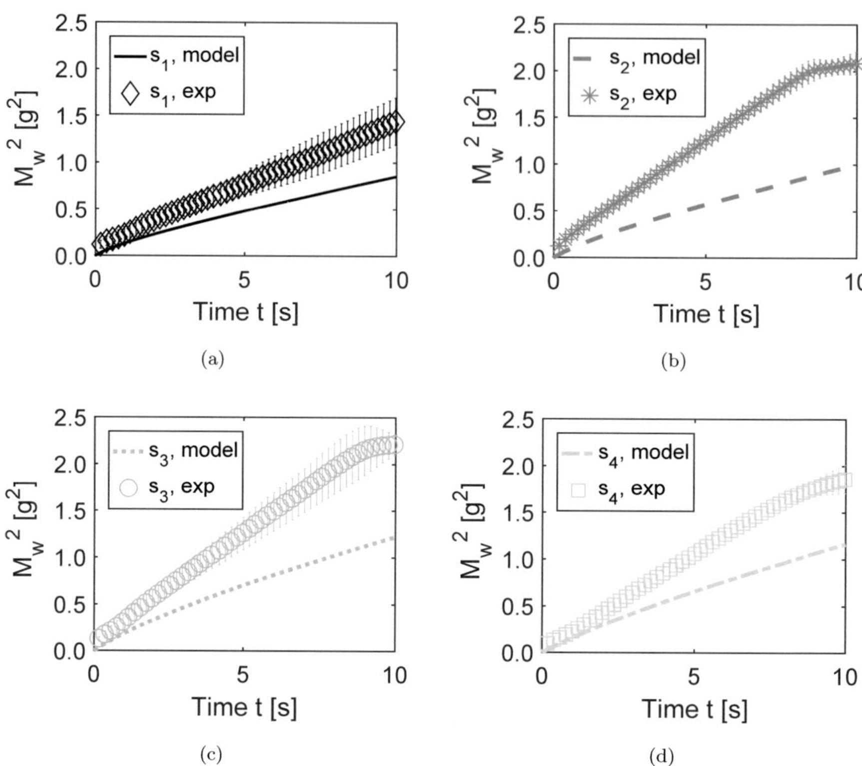

FIGURE A.2: Squared mass of water as function of time during penetration into four different size fractions of sodium chloride powder. Comparison of experiment and model for s_1 (a), s_2 (b), s_3 (c) and s_4(d).

Appendix B

Sucrose and hydrophobic glass

In Table B.1, the effective pore radii of mixtures containing sucrose and hydrophobic glass beads are presented. The values were calculated by means of the capillary constants and the porosities of the mixtures and the cross sectional area of the Washburn cylinder (Equation 4.5). The effective pore radii are required for the model calculations.

TABLE B.1: Effective pore radius of different mixtures of sucrose and hydrophobic glass beads.

Surface fraction sucrose f_1 [%]	Surface fraction hydrophobic glass f_2 [%]	Effective pore radius $r_{p,eff}$ [µm]
100	0	6.78
95	5	6.10
80	20	6.41
50	50	7.14

Appendix C

Sodium chloride and shellac coated glass

Figure C.1 shows the viscosity and density as a function of time predicted by the model for four different mixtures of sodium chloride and shellac coated glass beads. It can be observed that the viscosity and the density increase with time, while the effect of increasing hydrophobic content from 0 % to 50 % is not significant.

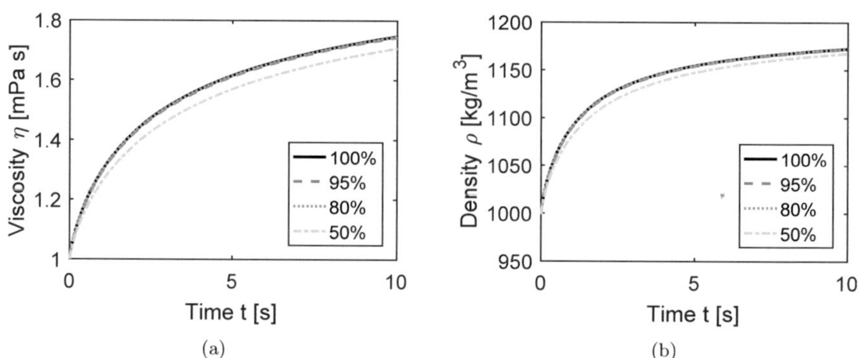

FIGURE C.1: Viscosity (a) and density (b) as a function of time for four different mixtures of sodium chloride and shellac coated glass beads predicted by the model

In Table C.1, the effective pore radii of mixtures containing sodium chloride and shellac coated glass beads are presented. The values were calculated by means of the capillary constants and the porosities of the mixtures and the cross sectional area of the Washburn cylinder (Equation 4.5). The effective pore radii were used for the model calculations.

TABLE C.1: Effective pore radius of different mixtures of sodium chloride and shellac coated glass beads.

Surface fraction sodium chloride f_1 [%]	Surface fraction shellac coated glass f_2 [%]	Effective pore radius $r_{p,eff}$ [µm]
100	0	4.65
95	5	5.01
80	20	6.03
50	50	7.80

Appendix D

Determination of change of pore network

In order to determine the change of pore network, the series of images in Figure 6.31 was used to calculate the change of pore radius, porosity and mean particle diameter. The equations D.1, D.2 and D.3 represent the alteration of the three pore network properties depending on time.

$$r_{p,eff} = 11.21e^{-9} \cdot t + 4.672e^{-6}. \tag{D.1}$$

$$\varepsilon = -4.510e^{-6} \cdot t^2 + 2.956e^{-3} \cdot t + 0.501. \tag{D.2}$$

$$d_{50} = -0.6912 \cdot t + 288. \tag{D.3}$$

Bibliography

Adhikari, B., Howes, T., Shrestha, A., and Bhandari, B. R. (2007). Effect of surface tension and viscosity on the surface stickiness of carbohydrate and protein solutions. *Journal of Food Engineering*, 79(4):1136–1143.

Anema, S. G., Lowe, E. K., and Li, Y. (2004). Effect of ph on the viscosity of heated reconstituted skim milk. *International Dairy Journal*, 14(6):541–548.

Asadi, M., editor (2007). *Beet-sugar handbook*. Wiley-Interscience, Hoboken, N.J.

Asioli, D., Aschemann-Witzel, J., Caputo, V., Vecchio, R., Annunziata, A., Næs, T., and Varela, P. (2017). Making sense of the "clean label" trends: A review of consumer food choice behavior and discussion of industry implications. *Food Research International*, 99:58–71.

Benavente, D., Lock, P., Ángeles García Del Cura, M., and Ordóñez, S. (2002). Predicting the capillary imbibition of porous rocks from microstructure. *Transport in Porous Media*, 49(1):59–76.

Bhandari, B., Bansal, N., Zhang, M., and Schuck, P. (2013). Handbook of food powders.

Bhandari, B. R., Senoussi, A., Dumoulin, E. D., and Lebert, A. (1993). Spray drying of concentrated fruit juices. *Drying Technology*, 11(5):1081–1092.

Bico, J. and Quere, D. (2002). Rise of liquids and bubbles in angular capillary tubes. *Journal of Colloid and Interface Science*, 247(1):162–166.

Birdi, K. S., Vu, D. T., and Winter, A. (1988). Capillary rise of liquids in rectangular tubings. *Colloid & Polymer Science*, 266(5):470–474.

Bosanquet, C. H. (1923). Lv. on the flow of liquids into capillary tubes. *The London, Edinburgh, and Dublin Philosophical Magazine and Journal of Science*, 45(267):525–531.

Bruil, H. G. and van Aartsen, J. J. (1974). The determination of contact angles of aqueous surfactant solutions on powders. *Colloid and Polymer Science*, 252(1):32–38.

Busscher, H. J., van Pelt, A., de Boer, P., de Jong, H. P., and Arends, J. (1984). The effect of surface roughening of polymers on measured contact angles of liquids. *Colloids and Surfaces*, 9(4):319–331.

Carman, P. C. (1956). *Flow of gases through porous media*. Academic Press.

Cassie, A. B. and Baxter, S. (1944). Wettability of porous surfaces. *Transactions of the Faraday society*, 40(1):546–551.

Chapuis, O. and Prat, M. (2007). Influence of wettability conditions on slow evaporation in two-dimensional porous media. *Physical Review E*, 75(4):046311.

Charles-Williams, H., Wengeler, R., Flore, K., Feise, H., Hounslow, M. J., and Salman, A. D. (2013). Granulation behaviour of increasingly hydrophobic mixtures. *Powder Technology*, 238:64–76.

Chau, T. T. (2009). A review of techniques for measurement of contact angles and their applicability on mineral surfaces. *Minerals Engineering*, 22(3):213–219.

Chebbi, R. (2007). Dynamics of liquid penetration into capillary tubes. *Journal of Colloid and Interface Science*, 315(1):255–260.

Chen, X. D. and Patel, K. C. (2008). Manufacturing better quality food powders from spray drying and subsequent treatments. *Drying Technology*, 26(11):1313–1318.

Cox, R. G. (1986). The dynamics of the spreading of liquids on a solid surface. part 1. viscous flow. *Journal of Fluid Mechanics*, 168(-1):169.

Crowley, S. V., Desautel, B., Gazi, I., Kelly, A. L., Huppertz, T., and O'Mahony, J. A. (2015). Rehydration characteristics of milk protein concentrate powders. *Journal of Food Engineering*, 149:105–113.

Cuq, B., Rondet, E., and Abecassis, J. (2011). Food powders engineering, between knowhow and science: Constraints, stakes and opportunities. *Powder Technology*, 208(2):244–251.

Czachor, H. (2006). Modelling the effect of pore structure and wetting angles on capillary rise in soils having different wettabilities. *Journal of Hydrology*, 328(3-4):604–613.

Dang-Vu, T. and Hupka, J. (2009). Characterization of porous materials by capillary rise method. *Phisicochemical Problems of Mineral Processing*, 39(1):47–65.

Darcy, H. (1856). *Les fontaines publiques de la ville de Dijon: exposition et application*. Victor Dalmont.

de Gennes, P.-G., Brochard-Wyart, F., and Quéré, D. (2004). *Capillarity and Wetting Phenomena.* Springer New York, New York, NY.

Dobrinski, P., Krakau, G., and Vogel, A. (2010). *Physik für Ingenieure.* Springer.

Dullien, F. A. L. (1992). *Porous media: fluid transport and pore structure.* Academic Press.

Dupas, J. (2012). *Wetting of soluble polymers.* Thesis, Université Pierre et Marie Curie - Paris VI.

Dupas, J., Forny, L., and Ramaioli, M. (2015). Powder wettability at a static air-water interface. *Journal of Colloid and Interface Science,* 448(1):51–56.

Dupas, J., Girard, V., and Forny, L. (2017). Reconstitution properties of sucrose and maltodextrins. *Langmuir : the ACS journal of surfaces and colloids,* 33(4):988–995.

Engasser, J.-M., Chamouleau, F., Chebil, L., and Ghoul, M. (2008). Kinetic modeling of glucose and fructose dissolution in 2-methyl 2-butanol. *Biochemical Engineering Journal,* 42(2):159–165.

Fang, Y., Selomulya, C., and Chen, X. D. (2007). On measurement of food powder reconstitution properties. *Drying Technology,* 26(1):3–14.

Fitzpatrick, J. J., van Lauwe, A., Coursol, M., O'Brien, A., Fitzpatrick, K. L., Ji, J., and Miao, S. (2016). Investigation of the rehydration behaviour of food powders by comparing the behaviour of twelve powders with different properties. *Powder Technology,* 297:340–348.

Forny, L., Marabi, A., and Palzer, S. (2011). Wetting, disintegration and dissolution of agglomerated water soluble powders. *Powder Technology,* 206(1-2):72–78.

Freudig, B., Hogekamp, S., and Schubert, H. (1999). Dispersion of powders in liquids in a stirred vessel. *Chemical Engineering and Processing: Process Intensification,* 38(4-6):525–532.

Fries, N. and Dreyer, M. (2008a). An analytic solution of capillary rise restrained by gravity. *Journal of Colloid and Interface Science,* 320(1):259–263.

Fries, N. and Dreyer, M. (2008b). The transition from inertial to viscous flow in capillary rise. *Journal of Colloid and Interface Science,* 327(1):125–128.

Gaiani, C., Ehrhardt, J. J., Scher, J., Hardy, J., Desobry, S., and Banon, S. (2006). Surface composition of dairy powders observed by x-ray photoelectron spectroscopy and effects on their rehydration properties. *Colloids and Surfaces B: Biointerfaces,* 49(1):71–78.

Gaiani, C., Morand, M., Sanchez, C., Arab Tehrany, E., Jacquot, M., Schuck, P., Jeantet, R., and Scher, J. (2010). How surface composition of high milk proteins powders is influenced by spray-drying temperature. *Colloids and Surfaces B: Biointerfaces*, 75(1):377–384.

Gaiani, C., Scher, J., Ehrhardt, J. J., Linder, M., Schuck, P., Desobry, S., and Banon, S. (2007). Relationships between dairy powder surface composition and wetting properties during storage: importance of residual lipids. *Journal of Agricultural and Food Chemistry*, 55(16):6561–6567.

Galet, L., Patry, S., and Dodds, J. (2010). Determination of the wettability of powders by the washburn capillary rise method with bed preparation by a centrifugal packing technique. *Journal of Colloid and Interface Science*, 346(2):470–475.

Galet, L., Vu, T. O., Oulahna, D., and Fages, J. (2004). The wetting behaviour and dispersion rate of cocoa powder in water. *Food and Bioproducts Processing*, 82(4):298–303.

Hammes, M. V., Heberle, E. S., da Silva, P. R., Noreña, C. P. Z., Englert, A. H., and Cardozo, N. S. M. (2018). Mathematical modeling of the capillary rise of liquids in partially soluble particle beds. *Powder Technology*, 325:21–30.

Hamraoui, A., Thuresson, K., Nylander, T., and Yaminsky, V. (2000). Can a dynamic contact angle be understood in terms of a friction coefficient? *Journal of Colloid and Interface Science*, 226(2):199–204.

Hapgood, K. P., Litster, J. D., Biggs, S. R., and Howes, T. (2002). Drop penetration into porous powder beds. *Journal of Colloid and Interface Science*, 253(2):353–366.

Hapgood, K. P., Litster, J. D., and Smith, R. (2003). Nucleation regime map for liquid bound granules. *AIChE Journal*, 49(2):350–361.

Harkins, W. D. and McLaughlin, H. M. (1925). The structure of films of water on salt solutions i. surface tension and adsorption for aqueous solutions of sodium chloride. *Journal of the American Chemical Society*, 47(8):2083–2089.

Hodges, G. E., Lowe, E. K., and Paterson, A. (1993). A mathematical model for lactose dissolution. *The Chemical Engineering Journal and the Biochemical Engineering Journal*, 53(2):B25–B33.

Hogekamp, S. and Schubert, H. (2003). Rehydration of food powders. *Food Science and Technology International*, 9(3):223–235.

Honig, P. (1953). *Principles of Sugar Technology*. Elsevier Science, Burlington.

Hussain, A. A., Abashar, M. E., and Al-Mutaz, I. S. (2006). Effect of ion sizes on separation characteristics of nanofiltration membrane systems. *Engineering Science*, 9:1–19.

Ichikawa, N. and Satoda, Y. (1994). Interface dynamics of capillary flow in a tube under negligible gravity condition. *Journal of Colloid and Interface Science*, 162(2):350–355.

Iveson, S. M., Holt, S., and Biggs, S. (2000). Contact angle measurements of iron ore powders. *Colloids and Surfaces A: Physicochemical and Engineering Aspects*, 166(1-3):203–214.

Iveson, S. M., Litster, J. D., Hapgood, K., and Ennis, B. J. (2001). Nucleation, growth and breakage phenomena in agitated wet granulation processes: A review. *Powder Technology*, 117(1-2):3–39.

Jeske, S., Zannini, E., and Arendt, E. K. (2018). Past, present and future: The strength of plant-based dairy substitutes based on gluten-free raw materials. *Food Research International*, 110:42–51.

Jinapong, N., Suphantharika, M., and Jamnong, P. (2008). Production of instant soymilk powders by ultrafiltration, spray drying and fluidized bed agglomeration. *Journal of Food Engineering*, 84(2):194–205.

Kammerhofer, J., Fries, L., Dupas, J., Forny, L., Heinrich, S., and Palzer, S. (2018a). Impact of hydrophobic surfaces on capillary wetting. *Powder Technology*, 328:367–374.

Kammerhofer, J., Fries, L., Dymala, T., Dupas, J., Forny, L., Heinrich, S., and Palzer, S. (2018b). Penetration rates into heterogeneous model systems and soluble food material. *Powder Technology*, 339:765–774.

Katoh, K., Wakimoto, T., and Nitta, S. (2010). A study on capillary flow under the effect of dynamic wetting. *Journal of the Japanese Society for Experimental Mechanics*, 10(Special Issue):s62–s66.

Katoh, K., Wakimoto, T., Yamamoto, Y., and Ito, T. (2015). Dynamic wetting behavior of a triple-phase contact line in several experimental systems. *Experimental Thermal and Fluid Science*, 60:354–360.

Kestin, J., Khalifa, H. E., and Correia, R. J. (1981). Tables of the dynamic and kinematic viscosity of aqueous nacl solutions. *Journal of Physical and Chemical Reference Data*, 10(1):71–88.

Kiesvaara, J. and Yliruusi, J. (1993). The use of the washburn method in determining the contact angles of lactose powder. *International Journal of Pharmaceutics*, 92(1-3):81–88.

Kilau, H. W. and Pahlman, J. E. (1987). Coal wetting ability of surfactant solutions and the effect of multivalent anion additions. *Colloids and Surfaces*, 26:217–242.

Kim, E. H.-J., Chen, X. D., and Pearce, D. (2002). Surface characterization of four industrial spray-dried dairy powders in relation to chemical composition, structure and wetting property. *Colloids and Surfaces B: Biointerfaces*, 26(3):197–212.

Kirdponpattara, S., Phisalaphong, M., and Newby, B. Z. (2013). Applicability of washburn capillary rise for determining contact angles of powders/porous materials. *Journal of Colloid and Interface Science*, 397:169–176.

Koiranen, T., Kilpiö, T., Nurmi, J., and Nordén, H. V. (1999). The modelling and simulation of dissolution of sucrose crystals. *Journal of Crystal Growth*, 198-199:749–753.

Kowalska, J. and Lenart, A. (2005). The influence of ingredients distribution on properties of agglomerated cocoa products. *Journal of Food Engineering*, 68(2):155–161.

Kozeny, J. (1927). Über kapillare leitung des wassers im boden, sitz. der wien. *Akad. der Wissenschaften*, 136.

Lerk, C. F., Schoonen, A., and Fell, J. T. (1976). Contact angles and wetting of pharmaceutical powders. *Journal of Pharmaceutical Sciences*, 65(6):843–847.

Li, D. and Neumann, A. W. (1992). Surface heterogeneity and contact angle hysteresis. *Colloid and Polymer Science*, 270(5):498–504.

Lide, D. R. and Kehiaian, H. V. (1994). *CRC handbook of thermophysical and thermochemical data*. CRC Press, Boca Raton.

Litster, J., Ennis, B., and Liu, L. (2004). *The science and engineering of granulation processes*, volume 15 of *Particle technology series*. Kluwer, Dordrecht.

Lucas, R. (1918). Rate of capillary ascension of liquids. *Kolloid Z*, 23(15):15–22.

Marmur, A. (1992a). Contact angle equilibrium: The intrinsic contact angle. *Journal of Adhesion Science and Technology*, 6(6):689–701.

Marmur, A. (1992b). Penetration and displacement in capillary systems of limited size. *Advances in Colloid and Interface Science*, 39:13–33.

Martic, G., Gentner, F., Seveno, D., Coulon, D., de Coninck, J., and Blake, T. D. (2002). A molecular dynamics simulation of capillary imbibition. *Langmuir*, 18(21):7971–7976.

Masoodi, R. and Pillai, K. M. (2012). *Wicking in porous materials: traditional and modern modeling approaches*. CRC Press.

Mathlouthi, M. and Reiser, P. (1995). *Sucrose: Properties and applications*. Blackie Academic & Professional, London.

McDonald, E. J. and Turcotte, A. L. (1948). Density and refractive indices of lactose solutions. *Journal of research of the National Bureau of Standards*, 41(1):63–68.

McSweeney, P. and Fox, P. F., editors (2009). *Advanced Dairy Chemistry: Lactose, Water, Salts and Minor Constituents*. Springer New York, New York, NY.

Middleman, S. (1995). *Modeling Axisymmetric Flows*. Academic Press, San Diego.

Miller-Chou, B. A. and Koenig, J. L. (2003). A review of polymer dissolution. *Progress in Polymer Science*, 28(8):1223–1270.

Mitchell, W. R., Forny, L., Althaus, T., Niederreiter, G., Palzer, S., Hounslow, M. J., and Salman, A. D. (2017). Surface tension effects in the reconstitution of food powders. *Proceedings, Granulation Workshop Sheffield*.

Mitchell, W. R., Forny, L., Althaus, T. O., Niederreiter, G., Palzer, S., Hounslow, M. J., and Salman, A. D. (2015). Mapping the rate-limiting regimes of food powder reconstitution in a standard mixing vessel. *Powder Technology*, 270(1):520–527.

Mohos, F. Á. (2017). Data on engineering properties of materials used and made by the confectionery industry. In *Confectionery and Chocolate Engineering*, pages 617–642. John Wiley & Sons, Ltd.

Murata, T. and Naka, A. (1983). A modified penetration rate method for measuring the wettability of coal powders. *Journal of Japan Oil Chemists' Society*, 32(9):498–502.

Nguyen, T., Shen, W., and Hapgood, K. (2009). Drop penetration time in heterogeneous powder beds. *Chemical Engineering Science*, 64(24):5210–5221.

North, J. (2017). Global consumer trends in store for 2018. *Food New Zealand*, 17(6):20.

O'Brien, W., Craig, R., and Peyton, F. (1968). Capillary penetration between dissimilar solids. *Journal of Colloid and Interface Science*, 26(4):500–508.

Ortega-Rivas, E., Juliano, P., and Yan, H. (2006). *Food powders: physical properties, processing, and functionality*. Springer Science & Business Media.

Palzer, S. (2000). *Anreichern und Benetzen von pulverförmigen Lebensmitteln mit Flüssigkeiten in diskontinuierlichen Mischaggregaten*. Dissertation, Technische Universität München, München.

Palzer, S., Hiebl, C., Sommer, K., and Lechner, H. (2001). Einfluss der rauigkeit einer feststoffoberfläche auf den kontaktwinkel. *Chemie Ingenieur Technik*, 73(8):1032–1038.

Parker, A., Vigouroux, F., and Reed, W. F. (2000). Dissolution kinetics of polymer powders. *AIChE Journal*, 46(7):1290–1299.

Patocka, G., Cervenkova, R., Narine, S., and Jelen, P. (2006). Rheological behaviour of dairy products as affected by soluble whey protein isolate. *International Dairy Journal*, 16(5):399–405.

Peppas, N. A., Wu, J. C., and von Meerwall, E. D. (1994). Mathematical modeling and experimental characterization of polymer dissolution. *Macromolecules*, 27(20):5626–5638.

Pfalzer, L., Bartusch, W., and Heiss, R. (1973). Untersuchungen über die physikalischen eigenschaften agglomerierter pulver. *Chemie Ingenieur Technik*, 45(8):510–516.

Pitzer, K. S., Peiper, J. C., and Busey, R. H. (1984). Thermodynamic properties of aqueous sodium chloride solutions. *Journal of Physical and Chemical Reference Data*, 13(1):1–102.

Prestidge, C. A. and Ralston, J. (1995). Contact angle studies of galena particles. *Journal of Colloid and Interface Science*, 172(2):302–310.

Quéré, D. (1997). Inertial capillarity. *Europhysics Letters (EPL)*, 39(5):533–538.

Raux, P. S., Cockenpot, H., Ramaioli, M., Quéré, D., and Clanet, C. (2013). Wicking in a powder. *Langmuir : the ACS journal of surfaces and colloids*, 29(11):3636–3644.

Reineccius, G. A. (2004). The spray drying of food flavors. *Drying Technology*, 22(6):1289–1324.

Reinke, S. K., Hauf, K., Vieira, J., Heinrich, S., and Palzer, S. (2015). Changes in contact angle providing evidence for surface alteration in multi-component solid foods. *Journal of Physics D: Applied Physics*, 48(46):464001.

Roman-Gutierrez, A., Sabathier, J., Guilbert, S., Galet, L., and Cuq, B. (2003). Characterization of the surface hydration properties of wheat flours and flour components by the measurement of contact angle. *Powder Technology*, 129(1–3):37–45.

Rosen, M. J. (1978). *Surfactants and interfacial phenomena*. Wiley, New York, N.Y.

Rumpf, H. (1958). Grundlagen und methoden des granulierens. *Chemie Ingenieur Technik - CIT*, 30(3):144–158.

Saguy, I. S., Marabi, A., and Wallach, R. (2005a). Liquid imbition during rehydration of dry porous foods. *Innovative Food Science & Emerging Technologies*, 6(1):37–43.

Saguy, I. S., Marabi, A., and Wallach, R. (2005b). New approach to model rehydration of dry food particulates utilizing principles of liquid transport in porous media. *Trends in Food Science & Technology*, 16(11):495–506.

Sawamura S., Egoshi, N., Setoguchi, Y., and Matsuo, H. (2007). Solubility of sodium chloride in water under high pressure. *Fluid Phase Equilibria*, 254(1):158–162.

Schubert, H. (1978). Optimierung der größe und porosität von instantagglomeraten in bezug auf eine schnelle durchfeuchtung. *Verfahrenstechnik*, 12(5):296–301.

Schubert, H. (1987). Food particle technology. part i: Properties of particles and particulate food systems. *Journal of Food Engineering*, 6(1):1–32.

Schubert, H. (1990). Instantisieren pulverförmiger lebensmittel. *Chemie Ingenieur Technik*, 62(11):892–906.

Schultz, S. G. and Solomon, A. K. (1961). Determination of the effective hydrodynamic radii of small molecules by viscometry. *The Journal of General Physiology*, 44(6):1189–1199.

Schwedt, G. (2010). *Zuckersüße Chemie*. Wiley-VCH Verlag GmbH & Co. KGaA, Weinheim, Germany.

Selomulya, C., Fang, Y., Bhandari, B., Bansal, N., Zhang, M., and Schuck, P. (2013). Food powder rehydration. *Handbook of Food Powders: Processes and Properties*, 379.

Siebold, A., Nardin, M., Schultz, J., Walliser, A., and Oppliger, M. (2000). Effect of dynamic contact angle on capillary rise phenomena. *Colloids and Surfaces A: Physicochemical and Engineering Aspects*, 161(1):81–87.

Siebold, A., Walliser, A., Nardin, M., Oppliger, M., and Schultz, J. (1997). Capillary rise for thermodynamic characterization of solid particle surface. *Journal of Colloid and Interface Science*, 186(1):60–70.

Silva, J. V. C. and O'Mahony, J. A. (2017). Flowability and wetting behaviour of milk protein ingredients as influenced by powder composition, particle size and microstructure. *International Journal of Dairy Technology*, 70(2):277–286.

Staples, T. L. and Shaffer, D. G. (2002). Wicking flow in irregular capillaries. *Colloids and Surfaces A: Physicochemical and Engineering Aspects*, 204(1-3):239–250.

Stevens, P., Gypen, L., and Jennen-Bartholomeussen, R. (1974). Wettability of powders. *Farmaceutish Tijdschrift Voor Belgie*, 51:150–155.

Susana, L., Campaci, F., and Santomaso, A. C. (2012). Wettability of mineral and metallic powders: Applicability and limitations of sessile drop method and washburn's technique. *Powder Technology*, 226:68–77.

Thurmond, V. L., Potter, R. W., and Clynne, M. A. (1984). The densities of saturated solutions of nacl and kcl. (84-253).

Tsotsas, E. and Mujumdar, A. S. (2008). Modern drying technology vol. 1 computational tools at different scales. *Drying Technology*, 26(6):812–814.

van Loo, E. J., Hoefkens, C., and Verbeke, W. (2017). Healthy, sustainable and plant-based eating: Perceived (mis)match and involvement-based consumer segments as targets for future policy. *Food Policy*, 69:46–57.

Vargaftik, N. B., Volkov, B. N., and Voljak, L. D. (1983). International tables of the surface tension of water. *Journal of Physical and Chemical Reference Data*, 12(3):817–820.

VDI, editor (2013). *VDI-Wärmeatlas*. Springer reference. Springer Vieweg, Berlin, 11., bearb. und erw. aufl. edition.

Walstra, P. (1999). *Dairy technology: Principles of milk properties and processes*, volume 90 of *Food science and technology*. Marcel Dekker, New York.

Wang, Q., Ellis, P. R., and Ross-Murphy, S. B. (2002). Dissolution kinetics of guar gum powders. i. methods for commercial polydisperse samples. *Carbohydrate Polymers*, 49(2):131–137.

Wang, Q., Ellis, P. R., and Ross-Murphy, S. B. (2008). Dissolution kinetics of water-soluble polymers: The guar gum paradigm. *Carbohydrate Polymers*, 74(3):519–526.

Wangler, J. and Kohlus, R. (2017). Dynamics of capillary wetting of biopolymer powders. *Chemical Engineering & Technology*, 40(9):1552–1560.

Washburn, E. W. (1921). The dynamics of capillary flow. *Physical review*, 17(3):273–283.

Wenzel, R. N. (1936). Resistance of solid surfaces to wetting by water. *Industrial & Engineering Chemistry*, 28(8):988–994.

Wenzel, R. N. (1949). Surface roughness and contact angle. *The Journal of Physical Chemistry*, 53(9):1466–1467.

White, L. R. (1982). Capillary rise in powders. *Journal of Colloid and Interface Science*, 90(2):536–538.

Wu, P., Zhang, H., Nikolov, A., and Wasan, D. (2016). Rise of the main meniscus in rectangular capillaries: Experiments and modeling. *Journal of Colloid and Interface Science*, 461:195–202.

Yang, R. Y., Yu, A. B., Choi, S. K., Coates, M. S., and Chan, H. K. (2008). Agglomeration of fine particles subjected to centripetal compaction. *Powder Technology*, 184(1):122–129.

Yekeler, M., Ulusoy, U., and Hiçyılmaz, C. (2004). Effect of particle shape and roughness of talc mineral ground by different mills on the wettability and floatability. *Powder Technology*, 140(1):68–78.

Young, T. (1805). An essay on the cohesion of fluids. *Philosophical Transactions of the Royal Society of London*, 95(0):65–87.

Zhmud, Tiberg, and Hallstensson (2000). Dynamics of capillary rise. *Journal of Colloid and Interface Science*, 228(2):263–269.

Zotarelli, M. F., Martins da Silva, V., Durigon, A., Dupas Hubinger, M., and Borges Laurindo, J. (2017). Production of mango powder by spray drying and cast-tape drying. *Powder Technology*, 305:447–454.

Zou, R. P. and Yu, A. B. (1996). Evaluation of the packing characteristics of mono-sized non-spherical particles. *Powder Technology*, 88(1):71–79.